AVIATION

E-8 JSTARS

Northrop Grumman's Joint Surveillance Target Attack Radar System

SÉRGIO SANTANA

Library of Congress Control Number: 2018958779

Type set in Impact/Minion Pro/Univers LT Std

ISBN: 978-0-7643-5667-4
Printed in China

Published by Schiffer Publishing, Ltd.
4880 Lower Valley Road
Atglen, PA 19310
Phone: (610) 593-1777; Fax: (610) 593-2002
E-mail: Info@schifferbooks.com
www.schifferbooks.com

Acknowledgments

This volume, intended to be the very first exclusively written about the E-8 JSTARS, would not be the same without the invaluable and patient help provided by Anderson Subtil, a friend of mine and very skilled graphic artist, who designed all of the book's artworks; Alexandre Galante, another friend of mine, for his prompt assistance on some occasions; Dianne Baumert-Moyik, media contact for Military Surveillance & Targeting at Northrop Grumman; Airman 1st Class Amanda Bodony, 116th Air Control Wing; and SMSgt. Roger Parsons, public affairs superintendent, Georgia Air National Guard. Special thanks to Robert Biondi, Schiffer senior editor, and Carey Massimini, Schiffer senior coordinator, for their full, patient support—a big thank you!

Dedication

To my dear mother, Elza, and all my family members and close friends. This book is also dedicated to the memory of Sir Stephen Hawking (1942–2018), a perennial inspiration who left many lessons for humanity and said, "Look up at the stars and not down at your feet."

Contents

Introduction

The Northrop Grumman E-8 JSTARS is a final product derived from an idea that appeared for the first time in 1958, when the US Army developed a radar installed on a telescopic mast designed to detect moving targets. It would be elevated above the tree level, execute a rapid scan of the surrounding scenery, and then retracts. It was called "Periodically Elevated Electronic Kibitzer" (PEEK for short, a "kibitzer" being someone who looks over a cardplayer's shoulder). But the project was canceled.

During the next decade the concept reemerged in two formats: (1) a Moving Target Indicator Side-Looking Airborne Radar (MTI-SLAR), the Motorola AN/APS-94, installed in the Grumman OV-1 Mohawk, which saw service in Vietnam, Europe, and South Korea, and (2) its version mounted in helicopters, initially designated ALARM (for Airborne Long-Range Alerting Radar for MTI, mounted on the Bell UH-1 Huey, which was successfully demonstrated in 1972, then having been renamed SOTAS (Standoff Target Acquisition System). It was expected to offer a constant coverage over the ground forces in contrast to the strip-like one seen when the system was attached to the Mohawk. The SOTAS concept eventually accumulated good results in various exercises during the mid-1970s, which led to two such systems being built and delivered to the US Army in Europe until 1979.

However, the US Air Force had been investigating the same capabilities for ten years, since 1969, and developed the MLRS (Multi-Lateration Radar Surveillance and Strike System), designed to detect airborne and ground targets. It was the first step toward a more ambitious program: the "Assault Breaker," later renamed "Pave Mover," developed in 1975 as a private venture funded by Norden and Grumman, which also included a target-destroying capability by means of "Battlefield Interdiction Missiles" linked to a Ground Moving Target Indicator and Synthetic Aperture Radar. That system was the result of a study by the Defense Science Board, with the final report titled *Summer Study of Conventional Counterforce against a Pact Attack* (Washington, DC: Office of the Secretary of Defense, 1976), which was dedicated to finding a way to neutralize the Warsaw Pact's numerical superiority in Europe and led the USAF to award development contracts to Hughes Aircraft and Grumman/Norden two years later for a system planned to support theater operations (in contrast to the SOTAS, limited to the Army Corps specifications).

But there was no sense in funding two similar systems for the same requirement, and the SOTAS was canceled in 1981, followed by the "Pave Mover." In May 1982, the Chiefs of Staff of the Air Force and Army agreed in a joint program, called JSTARS (**J**oint **S**urveillance **T**arget **A**ttack **R**adar **S**ystem), designed to serve both as a theater and tactical asset to meet Air Force and Army Corps requirements. In order to achieve that, it was decided the system would be split into an airborne segment and a ground one, under the responsibility of the Army Corps. To host the former, many options were evaluated, from the Boeing B-52 (which would have an offensive capability in addition to the surveillance system) to Lockheed C-130 and U-2/TR-1 and even the Grumman OV-10 Bronco. The C-18 (military designation of the Boeing 707) was finally selected as the best suitable platform, considering costs and needs to be satisfied. The ground segment was designated Common Ground Station, to be developed by the Army's Communications Electronics Command.

In September 1985, the Full Scale Development contract for the system was awarded to Grumman Aerospace Company, with the AN/APY-3 radar development being subcontracted to Norden Systems. The contract would have two phases: Full Scale Devel-

Pave Mover antenna. *Daderot*

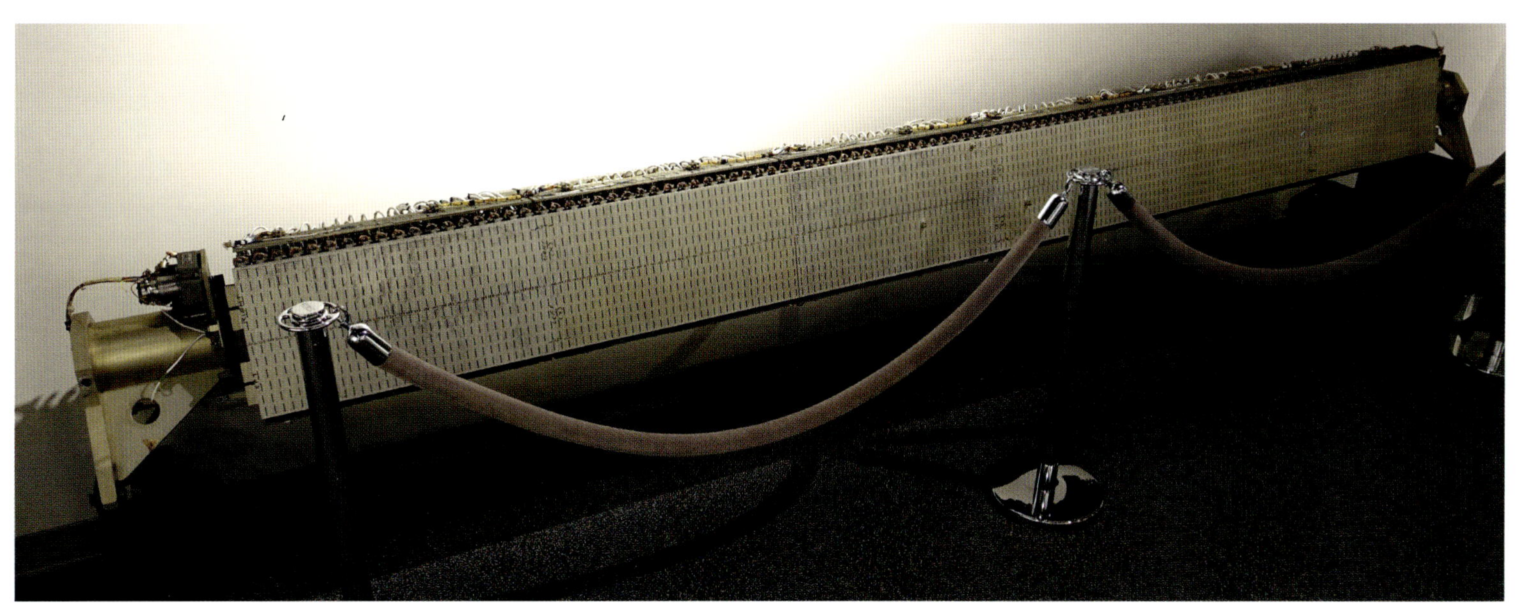

Another view of the Pave Mover antenna. *Daderot*

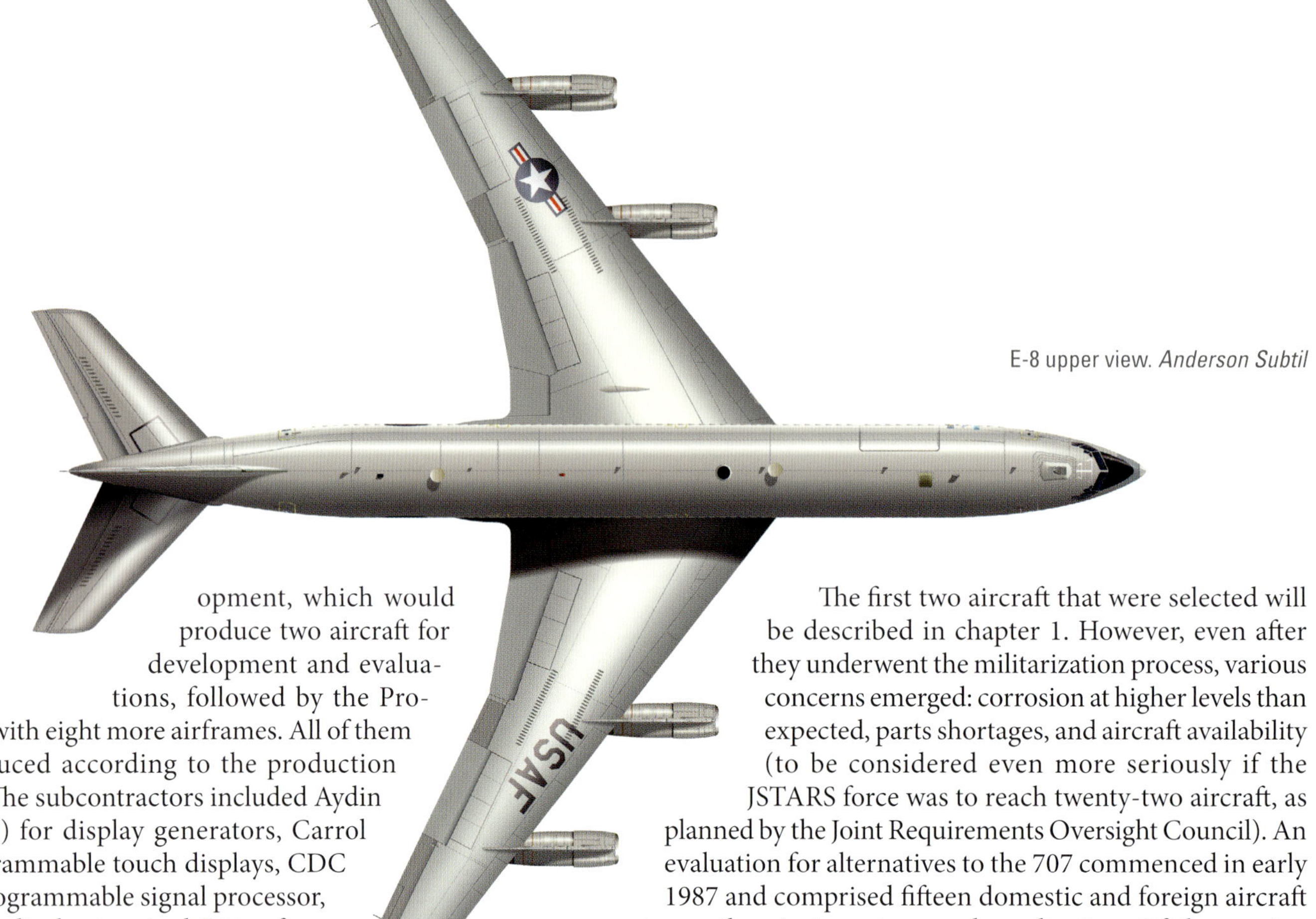

E-8 upper view. *Anderson Subtil*

opment, which would produce two aircraft for development and evaluations, followed by the Production Phase, with eight more airframes. All of them would be produced according to the production configuration. The subcontractors included Aydin (later Motorola) for display generators, Carrol Touch for programmable touch displays, CDC for the radar programmable signal processor, Hartman for the display terminal, Litton for the inertial measurement group, Magnavox for UHF radios, Miltope (later Data Metrics) for printers, Rolm (later Raytheon) for central data processors, and Telephonics for the intercommunication system.

The weight of the systems to be carried implied that the JSTARS airborne segment would be installed aboard the cargo version of the 707, with its stronger flooring compared to the one installed on the aircraft's passenger versions. The aircraft would have to depart from a 10,000-foot runway, with over 37,000 lbs. of mission equipment, and operate at an altitude between 34,000 and 42,000 ft. Its mission time, unrefueled, would be eleven hours at 450 kt. ground speed. It would be equipped with fifteen mission consoles and a mission crew of thirty-four specialists. Finally, the system would have to provide a 270kVA power with an engine out, in order to generate enough power and cooling.

The first two aircraft that were selected will be described in chapter 1. However, even after they underwent the militarization process, various concerns emerged: corrosion at higher levels than expected, parts shortages, and aircraft availability (to be considered even more seriously if the JSTARS force was to reach twenty-two aircraft, as planned by the Joint Requirements Oversight Council). An evaluation for alternatives to the 707 commenced in early 1987 and comprised fifteen domestic and foreign aircraft types, then in inventory and production. Of those, nine met the JSTARS technical requirements: Boeing E-6 TACAMO (a militarized 707 flown by the US Navy), B767-200, B747SP, B747-200, McDonnell Douglas DC-10-30 and MD-11, and Airbus A310-300 and A340-200.

After a brief period in which the E-6 was considered to be the best solution in terms of costs, availability, and risks, it was put aside on the grounds of the expensive costs required by Boeing to convert it into a JSTARS model. At the same time, the 707 assembly line was to be closed due to the lack of new orders for the type, which led to a new evaluation for a new alternative to replace the 707 for the JSTARS program. In addition to the types previously evaluated, this new round of reviews, which took place in 1989, also included the Boeing C-17A, Boeing KC-135/135R, B757-200, B767-300, Lockheed C-141B, and McDonnell-Douglas MD-87UHB. The radar performance in a given airframe remained the main criterion, followed by lack of physical support, lack of enough power,

E-8 frontal view. *Anderson Subtil*

E-8 assembly line.
Northrop Grumman

insufficient ground clearance, and radar beam blockage caused by engine nacelle placement. After all testing was completed, only three types were short listed and eliminated, after a careful examination: B767 (it would require dropping the AN/APY-3 radome farther below the fuselage, thus negatively affecting its aerodynamics in normal operational conditions; it also lacked an adequate electrical and environmental control system), A340 (the same issues as found in the B767), and MD-11 (the AN/APY-3 radome would have to be relocated too, but the type had better electrical and environmental systems than the previous candidates). It should be said that the MD-11 had better availability and affordability than the used B707s, but since the JSTARS system was already integrated on used 707s, this solution was declared as having the lowest technical and schedule risks in October 1989.

CHAPTER 1

The E-8A, E-8B, and E-8C

Well before those discussions, however, the JSTARS Joint Program Office (JPO) decided to select used Boeing 707s as the host platform for the system's airborne segment, but it imposed limitations on the number of hours and cycles (takeoffs and landings) at 50,000 and 20,000, respectively.

The JPO selected two B707s, taking into account their availability, configuration, price, airworthiness, and FAA certification. Aircraft 1, B707-338C (c/n 19626-703, USAF serial number 86-0416), whose first civilian operator had been Qantas, had logged 51,647 hours and 20,474 cycles, while Aircraft 2, B707-323C (c/n 19574, ex–American Airlines, USAF serial number 86-0417) had accumulated 40,008 hours and 17,191 cycles. Aircraft 1 took to the air for radar test flight on December 22, 1988, although it had already been flown without its ventral radar housing eight months earlier, on April 1. In order to hide their real purposes, the aircraft received a camouflage painting scheme and were designated "C-18" and then "EC-18C."

Prior to that stage, however, both aircraft faced an intricate modification process that included the main landing-gear support structure, a crease beam, and a ram air and ground service inlet. An in-flight refueling receptacle, a vapor cycle machine, a new auxiliary power unit, and antennas were also installed. The first production delivery was scheduled to occur in August 1991, but in September 1990 the E-8As flew during "Operation Deep Strike," a NATO forces exercise in which the aircraft detected a Canadian armored convoy emulating a Soviet formation. Fifty-one "kills" out of seventy-five vehicles were logged by strike helicopters.

They had long-range navigation equipment, an automatic direction finder, weather radar, distance-measuring equipment with VHF omnidirectional range, a precision approach and landing system, a Litton Inertial Navigation System, a Telephonics intercom system, and a Rockwell-Collins flight management system, while the mission equipment package included five Raytheon model 920/866 computers, three Computing Devices International signal processors, ten Raytheon model 920 workstations, Miltope printers, a Cubic Defense Systems surveillance and control data link, F1613/A VHF radios with KY-58-encrypted terminals for the SINCGARS system, AN/ARC-164 UHF radios (also with KY-58), HF AN/ARC-190 radios, and HF JTIDS class II and SCDL terminals, both with KGV-8 crypto device.

The E-8B was originally the proposed production version. It was planned to have a newly built airframe, F108 turbofan engines, and fifteen operator consoles, but the series was limited to just one example, YE-8B, 88-0322, which was flown on June 12, 1990, without mission avionics. The E-8B program was terminated due to the decision to use refurbished 707s. Soon after that, it was stored at Davis-Monthan AFB and now, under serial number 1902, serves with the Royal Saudi Air Force as one of its E-3A Sentry.

The E-8C is the definitive production version, featuring eighteen Raytheon model 920 workstations (seventeen for operators and one for navigation/self-defense roles, each with Orbit International workstation and Interstate Electronic graphic displays), mission computers replaced by two Compaq Clippers from 2001, and Data Metrics printers from the fifth example. It has the same basic equipment as the E-8A, including a satellite communications link plus three F1613/A VHF radios with KY-58-encrypted terminals for the SINCGARS system, twelve AN/ARC-164 UHF radios (also with KY-58), two HF AN/ARC-190 radios, and two HF JTIDS class II and SCDL terminals, both with KGV-8 crypto device.

Sequence of frames taken from a Northrop Grumman Corporation video of the E-8A's first flight in Melbourne, Florida, on April 1, 1988. *Anderson Subtil and Alexandre Galante*

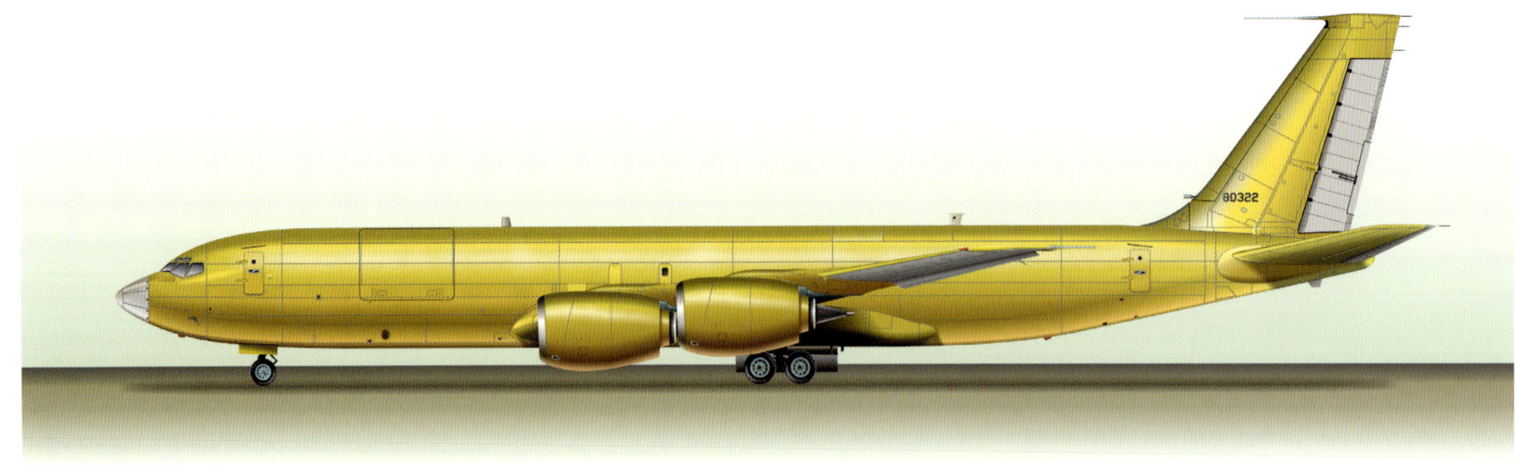

A side view of the canceled E-8B. *Anderson Subtil*

The E-8C in its definitive painting scheme. *Anderson Subtil*

An official meeting to celebrate the induction of the final E-8C, March 2005. *USAF*

The E-8C has been constantly upgraded in blocks. The US Air Force took delivery of the first E-8C, serial number 92-3289, on March 22, 1996, while the seventeenth and last example, an EC-18B, serial number 81-0896, named "Pride of Lake Charles," was handed over to the USAF on March 22, 2005.

E-8C Block 0

Four examples were built for test and evaluation roles and were later upgraded to the Block 20 version.

E-8C Block 10

This is the baseline configuration, with full interoperability with other US and Allied assets. It features TADIL-J secure communications, digital autopilot, and Y2K compliance. Six examples were modified, also being upgraded to the Block 20 version.

E-8C Block 20

This was modified to accept commercial off-the-shelf hardware, networked communications, US Army Improved Data Modem, and basic SINCGARS; ten aircraft were planned to be modified accordingly, with seven delivered between 2001 and 2003 and the remaining three until 2005. In addition to the systems above, it has a Goodrich MS-177 multispectral camera and AN/APY-7 radar. Currently, sixteen E-8C Block 20s are operated by USAF.

A fine study of the final E-8C. *USAF*

An E-8C JSTARS aircraft takes off for an intelligence, surveillance, and reconnaissance mission. *USAF file photo*

Addressing the crowd at Northrop Grumman's Melbourne, Florida, facility, Joint STARS program vice president Dale Burton leads the unveiling of the US Air Force's E-8C test bed aircraft, equipped with more efficient and effective engines. Unfortunately, the program did not go ahead, due to the lack of funds.
Northrop Grumman

A JSTARS flies over its home base.
Northrop Grumman

Northrop Grumman celebrates the thirty-year anniversary of the first JSTARS flight. *Northrop Grumman*

	Northrop Grumman E-8C JSTARS
Wingspan	145 feet, 9 inches
Length	152 feet, 11 inches
Height	42 feet, 6 inches
Maximum weight	336,000 pounds
Power plant	four Pratt & Whitney TF33-102C
Thrust	19,200 pounds each engine
Max. speed	644 mph
Service ceiling	42,000 feet
Range	over 2,800 miles
Crew	21

CHAPTER 2

Operational and Technical Description of the E-8C JSTARS

The E-8 is usually crewed by twenty-one persons, including a pilot, a copilot, a flight engineer, a navigator, a mission crew commander (MCC), a deputy mission crew commander (DMCC), two airborne target surveillance supervisors (ATSC), an airborne intelligence officer/technician (AIO/AIT), a senior director (SD), a senior director technician (SDT), two airborne operation technicians (AOT), a sensor management officer (SMO), two weapons directors (WD), two communication system technicians (CST), and three airborne missions system specialists (AMSS). For longer missions, this crew is eventually augmented by twenty-eight operators and six flight crew.

The crew-specific duties are as follows: the MCC is the responsible authority for assigned battle management-command, control, intelligence surveillance, and reconnaissance mission tasks and coordinates with the aircraft commander to ensure effective sortie and mission accomplishment. In addition to that, he/she also supervises execution of Higher Headquarters–assigned tasks, ensures crew member adherence to Rules of Engagement and Special Instructions, declares operations as normal/on-station/off-station and advises external agencies about the aircraft status, collates and compiles mission reports and summaries, has responsibility for accounting and safeguarding of classified materials and proper destruction, and tailors mission crew and positional responsibilities on the basis of mission requirements and operations. Finally, during decentralized operations, the MCC is the onboard authority for determining mission tasking.

The DMCC has the following responsibilities: acts as Army liaison to MCC and mission crew, ensures that the ground commander's intent is understood and that JSTARS crew members understand how ground operations will be executed, ensures that the ground commander and tactical ground stations are aware of on-station/off-station and aircraft status, manages information flow to supported ground units via radios and all available data links (Force XXI Battle Command Brigade-and-Below, Improved Data Modem, Surveillance Control Data Link, DATASAT, and AIRNET/INMARSAT), and coordinates with the on-ground fire support officer when required.

The AIO/AIT has to analyze incoming reports from external intelligence collection agencies and determine the impact on mission execution; ensure that amplifying intelligence data are fused as applicable to enhance the battle management-command, control, intelligence, surveillance, and reconnaissance mission; verify and update the order of battle data; operate the Integrated Broadcast Service; operate the automated information system; and report radar tracks both internally and externally to intelligence collection agencies for further collection and amplification.

For its turn, the SD monitors and assesses the current air/ground situation; coordinates mission changes with appropriate agencies; directs the execution of the battle management-command and control mission (which includes procedural control, managing mission changes, striking targets, and directing battle-space logistical efforts) with regard to Find, Fix, Track, Target, Engage, and Assess (F2T2EA); coordinates with the SMO for radar management and surveillance operations; and develops an effective communications plan.

The SMO conducts effective radar timeline management, informs crew of sensor anomalies, coordinates with SD for management of the Operations Section, and carries the SMO Flyaway Kit with specific documentation.

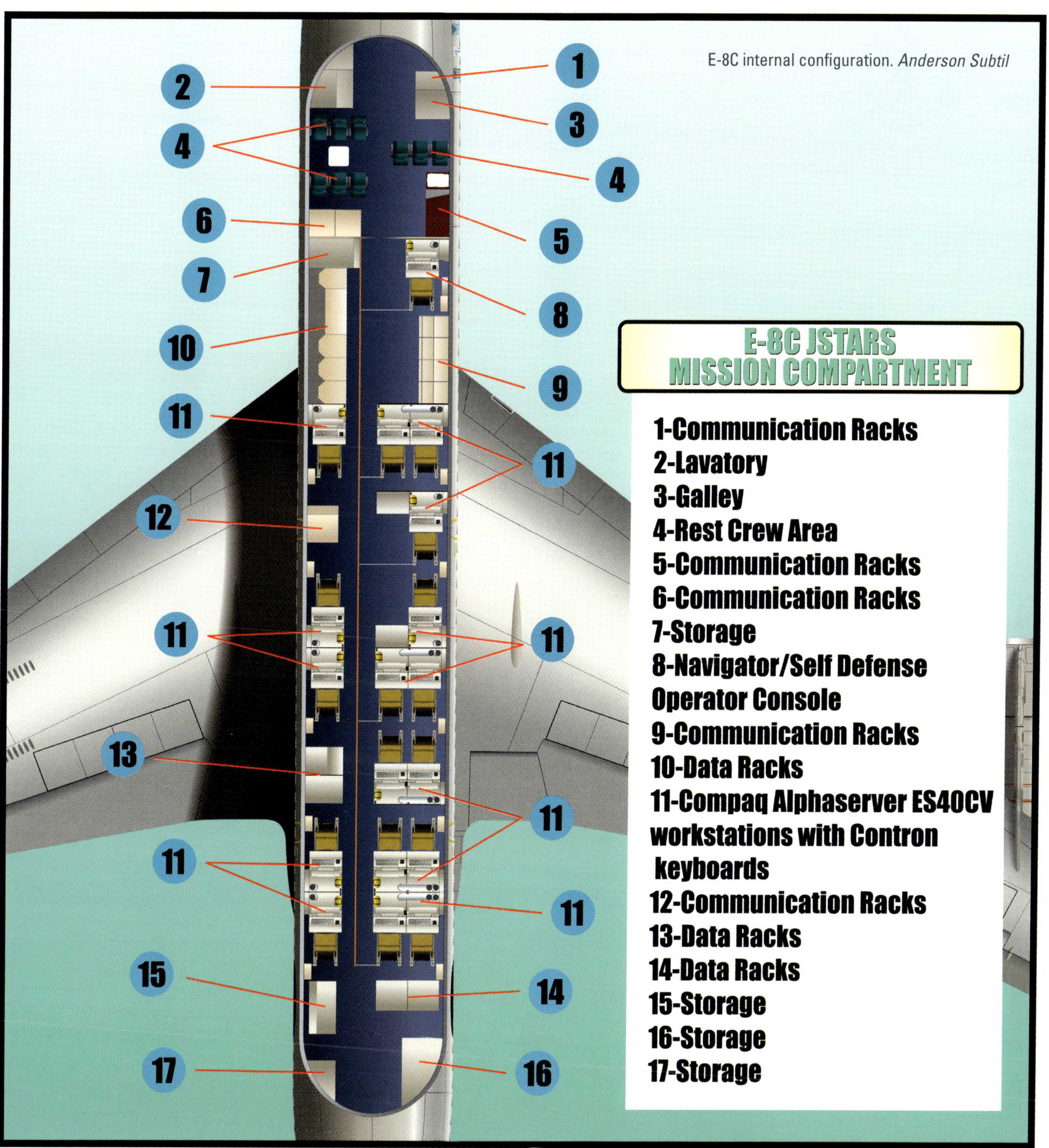

E-8C internal configuration. *Anderson Subtil*

The SSM ensures tracking responsibilities/continuity in the Area of Responsibility (AOR), coordinates with the CST for Joint Tactical Information Datalink System link operations, and oversees activities of the Surveillance Section.

The AWO conducts the battle management-command and control mission execution with regard to F2T2EA by using procedural control, target engagement, and TAC (A) managing ATO/ACO changes and directing battle-space logistical efforts.

The AOT uses use sensor data for accurate tracking in the assigned AOR.

The airborne target surveillance supervisor maintains voice and Surveillance Control Data Link (SCDL) contact with common ground stations to accomplish ground component commander objectives, as well as processing radar service requests as required.

The ART initiates, operates, and maintains radar and O&C (computer) systems. He or she monitors system status and troubleshoots malfunctions to keep systems operational, and acts as primary firefighter for emergencies involving these systems.

The CST initiates, operates, and maintains all aircraft communications, including voice and data link systems. He or she monitors system status and troubleshoots malfunctions to keep systems operational, as well as acting as primary firefighter for emergencies involving these systems.

Finally, liaison personnel from other US services and agencies are allowed to occupy mission crew seats and operate mission equipment, provided they are monitored by qualified JSTARS mission crew members.

The minimum flight crew is specified as a qualified aircraft commander, copilot, and flight engineer, while an instructor pilot with an "unqualified pilot" satisfies the two-pilot requirement. The minimum crew required to initialize and operate the mission system includes a navigator, an MCC, an SMO, two airborne radar technicians, and two communications system technicians.

The maximum flight duty for a minimum flight crew is twelve hours, while for an augmented crew (i.e., a qualified aircraft commander, navigator, and flight engineer in addition to the normal flight crew), it is sixteen hours.

The aircrew flight equipment consists of life preservers in the form of twenty-five MA-1 portable oxygen bottles attached to each firefighter smoke mask. These bottles are augmented by Emergency Personal Oxygen System (EPOS) kits, which provide a tertiary source of oxygen in the event that system oxygen and MA-1 portable oxygen are not available during an emergency requiring the use of supplemental oxygen. Each kit contains eight EPOS units. Sorties with more than thirty personnel require an EPOS kit onboard.

Regarding the E-8's operational employment, there are two types of alerts: the ALPHA alert and BRAVO alert. During the former, the aircrew must be capable of launching within one hour of crew notification of launch order. Crews should be quartered near the alert aircraft, with sufficient transportation to get them to the aircraft in minimum time. A crew will not stay on ALPHA alert duty for more than forty-eight hours. After forty-eight hours, the crew must be launched, released, or entered into predeparture crew rest. Crew duty day begins when the crew is notified of the launch order.

Under BRAVO alert, the aircrew is capable of launching within four hours of crew notification of launch order. Crew members are given twelve hours of pre-alert crew rest. After crew rest they are placed on telephone standby. A crew will not stay on BRAVO alert duty for more than forty-eight hours. After forty-eight hours, the crew must be launched, released, or entered into predeparture crew rest. Crew duty day begins when the crew is alerted for duty.

Mission development/planning is directed by the squadron operations officer (DO) or deployed detachment commander. The responsibility for mission planning is shared between the aircraft and mission crew commanders, the designated mission-planning cell, or both. Together they will ensure that the JSTARS flight profile is adequately planned and the crew is properly prepared for mission tasking and execution. The detachment commander may delegate authority to another officer under his or her authority.

Because intelligence is at the core of the E-8C's missions, the related tasks are coordinated by a senior intelligence officer. He or she is responsible for, among other things, leading, organizing, training, and equipping intelligence personnel and functions to support the unit mission as well as to solicit feedback from wing group and subordinate commanders to improve intelligence support processes. This specialized officer is aided by intelligence personnel who integrate with wing group weapons and tactics to fulfill necessary intelligence requirements during mission planning. He or she also participates in mission-planning teams, supporting the mission-planning cell by supplying materials and information to execute missions and satisfy tasking orders. That officer and his or her team also provide support to the mission-planning team to include information on extract mission employment from the appropriate tasking documents (e.g., air-tasking order; airspace control order; Reconnaissance, Surveillance, and Targeting Acquisition Annex; and special instructions), as well as to apply data to mission-planning processes and extract mission tasks and locations, reporting information from the collection deck.

Other tasks attributed to that team include determining the named area-of-interest descriptions and their significance and plot;

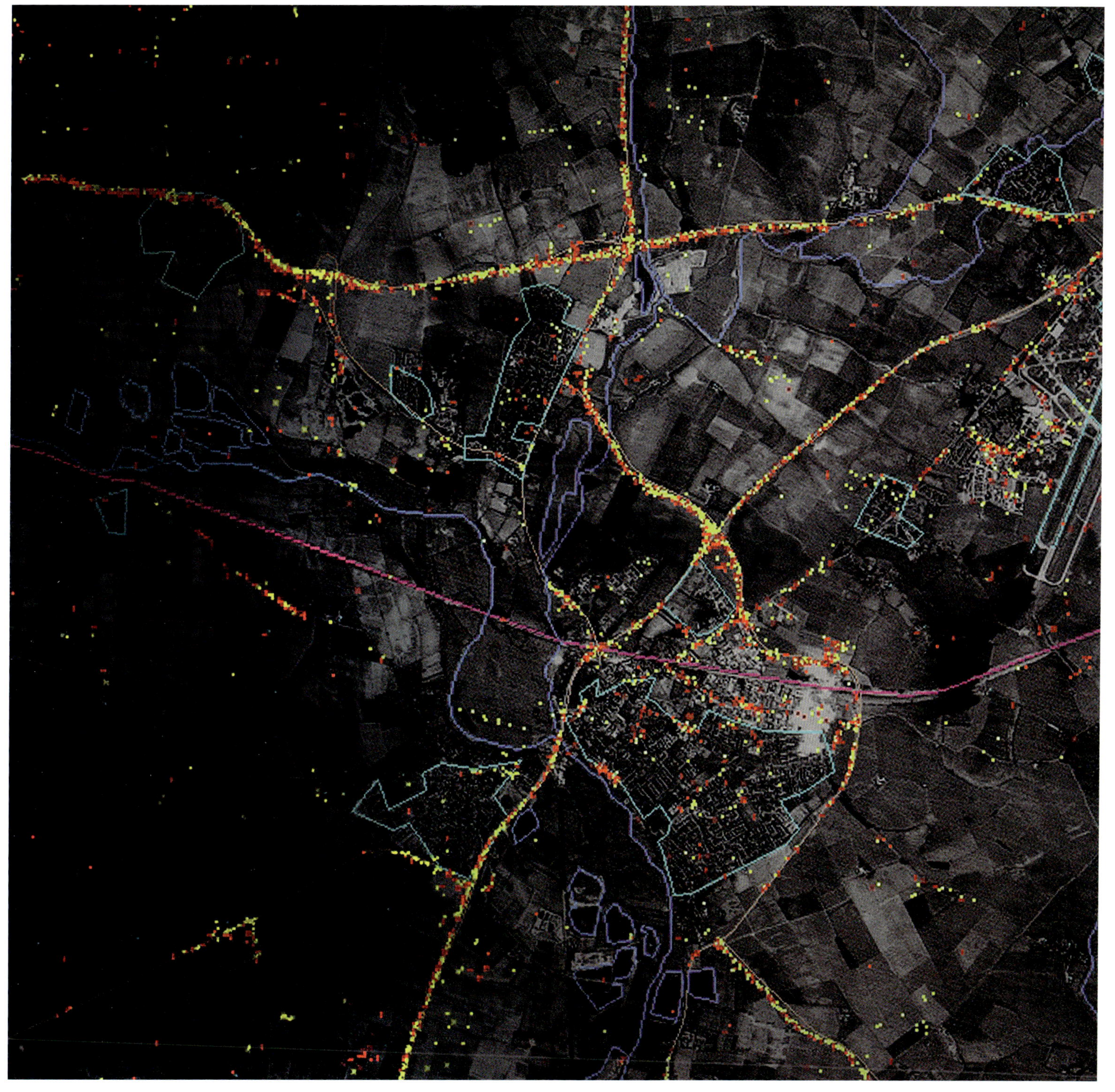

E-8C radar screen, GMTI mode. *US Department of Defense*

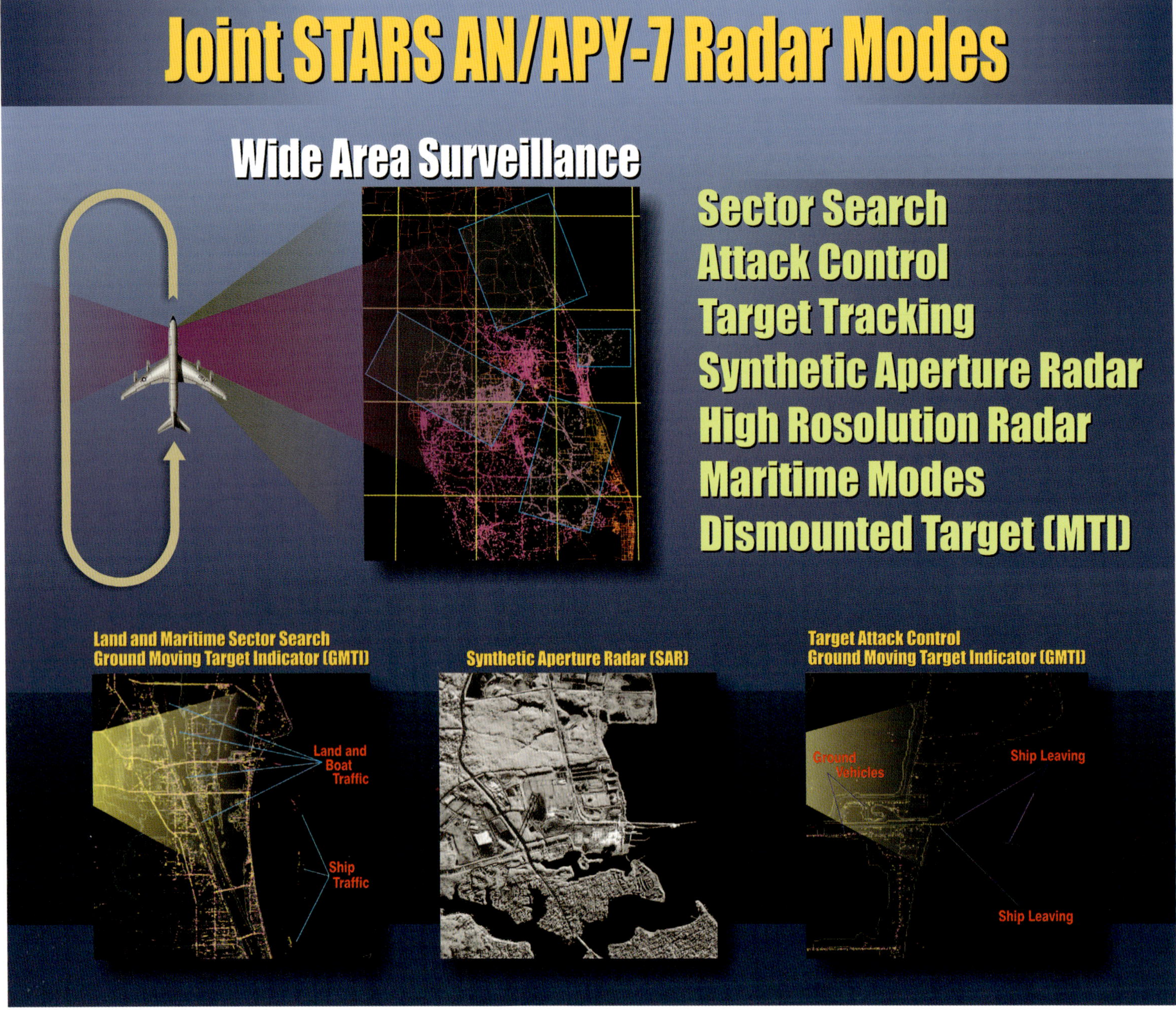

AN/APY-7 radar modes. *Northrop Grumman*

retrieving automated imagery and GMTI products by using the best available source; using intelligence systems and intelligence preparation of the operational-environment techniques to analyze threats and terrain affecting ingress, egress, and operating areas; providing expertise and analysis on potential threats and terrain affecting the mission; making en route and mission area charts as applicable; highlighting any friendly-force locations in mission areas; considering tenets of law of armed conflict and the rules of engagement applicable to the mission, coordinated with aircrew flight equipment and Survival, Evasion, Resistance, and Escape in order to provide the mission-planning team with personnel recovery procedures and related information; providing the mission-planning briefing, using the proper checklist; assisting the commander chief of the mission-planning cell in developing procedures to replan missions on the basis of new threats, Air Task Order changes, diverted missions, or a combination of these; and, last, updating preplanned missions to reflect the latest available intelligence information affecting the mission.

Finally, the same team is also charged with the debriefing of all aircrew on perishable, critical information of intelligence value prior to all other debriefings, disseminating critical debriefing information rapidly. The team also prepares appropriate sections of the debrief form/checklist (e.g., mission number, call sign) prior to mission return to base.

To be formally declared operational, an E-8C aircrew must be subjected to comprehensive training composed of the following phases: qualification training, transition or requalification training, continuation training, and upgrade training, the last of which provides training necessary to initially qualify aircrew in a basic position and flying duties without regard to a specific unit or squadron mission. Upon completion of these phases, the crew member reaches "basic aircraft qualification" status, after he or she has completed a flight evaluation and has been qualified to perform basic aircrew duties in the E-8. This status is maintained until the squadron commander or squadron operation officer confers a Combat Mission Ready or Basic Mission Capable certification.

The members of the flight crew are required to have a specific number of flight hours prior to be accepted into the E-8 JSTARS conversion training. A pilot must have a minimum of 1,000 hours of flight time to be qualified as an aircraft commander; a person who has logged fewer than 1,000 flight hours will be designated as a copilot.

Flight engineers with fewer than 500 flight hours must attend the E-8C Qualification Training course, while those with more than 500 flight hours may attend the E-8C Transition Training course. A combat systems officer with more than 750 hours may attend the E-8C Transition Training course. All other combat systems officers will attend the E-8C Qualification Training course.

To be designated a senior director, air battle managers with more than 750 total rated hours or one year as a ground theater air control system senior director may attend the E-8C Senior Director course. All other air battle managers will attend the E-8C Air Weapons Officer course.

To be designated a mission crew commander, air battle managers with more than 750 total flight hours or one year as a ground theater air control system senior director or higher may attend the E-8C Mission Crew Commander course. The air weapons officer does not have additional requirements.

Army members aboard the E-8 JSTARS also must have specific qualifications. In order to be designated as a deputy mission crew commander, for example, the trainee must be a warrant officer, captain, major, or lieutenant colonel.

Members entering the intel crew position of Airborne Intelligence (airborne intelligence officer or airborne intelligence technician) must be a graduate of an Air Force–accredited intelligence course.

In addition to the specific tasks of their positions, future crew members of the E-8 JSTARS are also trained in aircrew flight equipment, combat survival, local-area survival, emergency egress training, nonejection seats, water survival, crew resource management, and aircraft marshaling. This training must be completed within 120 days of the first flight (240 days for traditional Air National Guard [ANG] personnel).

JSTARS's current core mission equipment is the Norden Systems AN/APY-7, I-Band multimode radar, installed inside a 26-foot-long canoe beneath the aircraft forward fuselage.

The antenna can be tilted to either side of the aircraft, where it can develop a 120-degree field of view covering nearly 19,300 square miles. It is capable of detecting targets more than 150 miles away and can simultaneously track up to 600 targets. The radar also provides a limited capability to detect helicopters, rotating antennas, and slow-moving fixed-wing aircraft.

The AN/APY-7 operates in two basic modes: moving target indication (MTI) and synthetic aperture radar / fixed target indicator (SAR/FTI).

Both are activated to meet the requirements described in the Radar Service Requests (RSRs), which are requests by supported agencies for the radar to service a specific area and are executed by the radar on the basis of priorities established by the crew during mission planning. They are employed simultaneously. The radar sweeps the requested areas and provides a radar picture to the operators onboard the aircraft in the form of MTI. The frequency at which the radar sweeps a given area is based on the priority of the RSR and the requested revisit rate. Normally, the higher priority the RSR is, the higher the requested revisit rate will be, and the more time the radar will spend searching in that area.

When the MTI mode is selected, the radar detects moving vehicles on the basis of the Doppler shift return from wheeled vehicles and a double Doppler shift from a tracked vehicle, because its tracks are moving twice as fast as the vehicle itself and, therefore, create a double Doppler shift, which indicates such a vehicle. This Doppler and double Doppler shift is represented on the operator's console as magenta and yellow dots, respectively, which differentiates wheeled from tracked vehicles. However, because the E-8C radar is not designed to identify a contact as friendly or hostile, its operator needs to have cross-cued the radar picture with another sensor able to make these tasks.

The MTI mode detects moving vehicles, rotating antennas, and slow-moving aircraft. The targets are presented as dots moving at a given speed on the ground, and their location can be selected either in Universal Transverse Mercator (UTM) or latitudinal and longitudinal coordinates.

The MTI mode has several "submodes": (1) Ground reference coverage area, or GRCA, which is wide-area surveillance (WAS), low-resolution and usually low-priority RSR defined by a search area that the radar will attempt to continually keep in its field of view regardless of its position in the orbit. It is usually selected to cover up to a 200-square-mile area. A standard request for a revisit rate on a GRCA is 60 seconds. (2) Radar reference coverage area (RRCA), which is also a WAS, low-resolution, low-priority RSR. However, unlike the GRCA, its search area is defined by a fixed azimuth off the aircraft wing and does not have a defined search area on the ground. This submode is usually selected to validate the radar as operational en route to the area of operation. (3) Sector search, or SS, which provides a higher resolution and a quicker revisit rate (30 seconds) for a smaller area (18 x 18 miles) than the previous mode, delivering a more accurate and timely view of the vehicle activity in the defined area. Its priority is usually higher than that of the GRCA and provides the most accurate and timely MTI picture, thus being selected to aid the crew members in targeting. (4) Attack control, or AC, which is a high-resolution, high-priority, high-revisit rate (usually 10 to 12 seconds) MTI RSR. The AC mode is frequently requested during attack support operations and during ISR missions to aid in verifying possible vehicle movement. (5) Attack planning, or AP, which is similar to the AC in functionality but normally with a 15-to-20-second revisit rate over an area of 7 x 7 miles and lower priority. (6) Small area target classification (SATC), which differentiates between wheeled and tracked vehicles.

For its turn, the SAR mode is a high-resolution RSR, which means that the radar will spend a large amount of radar timeline, relative to the other RSRs, looking at a specific point on the ground, when the radar dwells (usually 2 to 3 seconds) on a specific area (usually 1.24 by 2.48 miles) requested by the operator and creates a radar image or picture of it. The SAR is the highest-priority RSR, being most frequently used in a change detection role, but it may also function in a very limited role in battle damage assessment. This mode can produce a still radar image of a given target, installation, or piece of terrain and is selected to locate those moving targets that have become stationary and are suspected to be in a given area. Stationary targets are detected and registered in their geographic position in the SAR image. This SAR function is usually cued by activity previously seen in the MTI mode. The targeted vehicles are automatically detected and highlighted. A SAR image looks like a black-and-white picture negative, with highly radar-reflective surfaces showing up as bright spots. It also allows limited target damage assessment (TDA) in addition to its targeting application. Its images do not allow determining if a vehicle has been destroyed, only that it is stationary. It can show damage to large man-made structures such as bridges.

A SAR can also be taken in the fixed target indicator (FTI) mode. This provides red dots overlaid on the SAR, which indicates the areas from which the radar received the greatest returns from the ground. The SAR-FTI can detect buildings, stationary vehicles, or assembly areas and is usually regarded as being able to provide better situational awareness of the ground picture. The targets are more easily recognized because in the FTI mode the target dot is presented on the screen without any surrounding terrain.

While operating in this mode, the radar had a range of 94.4 nautical miles. The interleaving capability of the radar allows the system to perform multiple operations (such as MTI, SS, and SAR) without disrupting the WAS revisit rate (unless many operations are being conducted simultaneously).

After some years of operation, the AN/APY-3, which was the original mission radar that equipped the E-8 JSTARS, was replaced by its AESA (active electronically scanned array, meaning it has electronic scan in azimuth and elevation) variant, designated AN/APY-7. In addition to the operational modes described above, it also has the following modes: enhanced land maritime mode, air moving-target indicator (for detecting low, slow-flying aircraft), aided target recognition, enhanced synthetic aperture radar, target

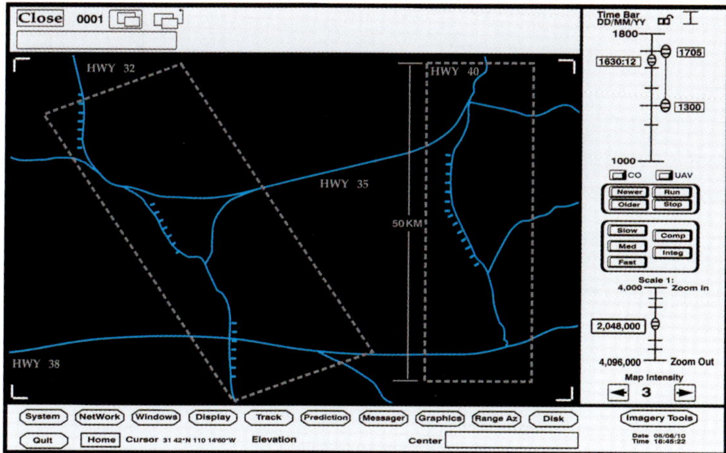

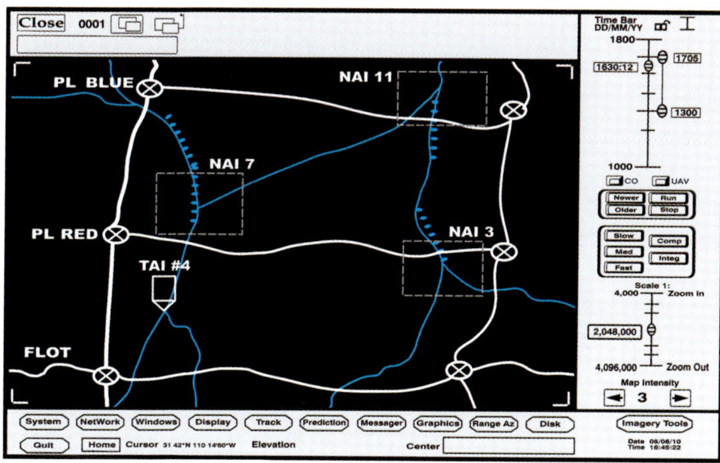

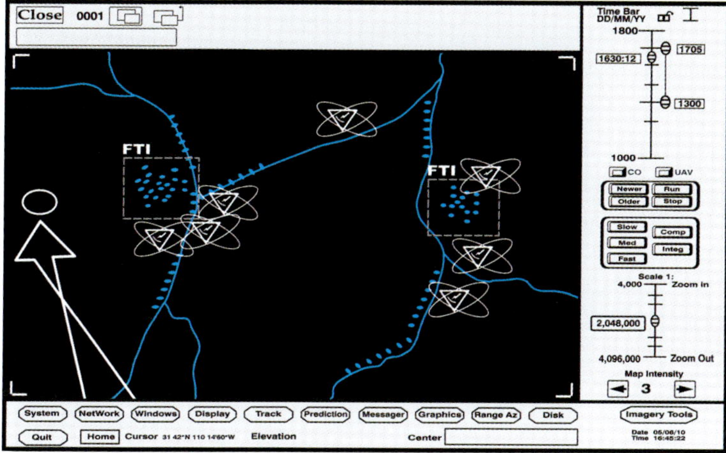

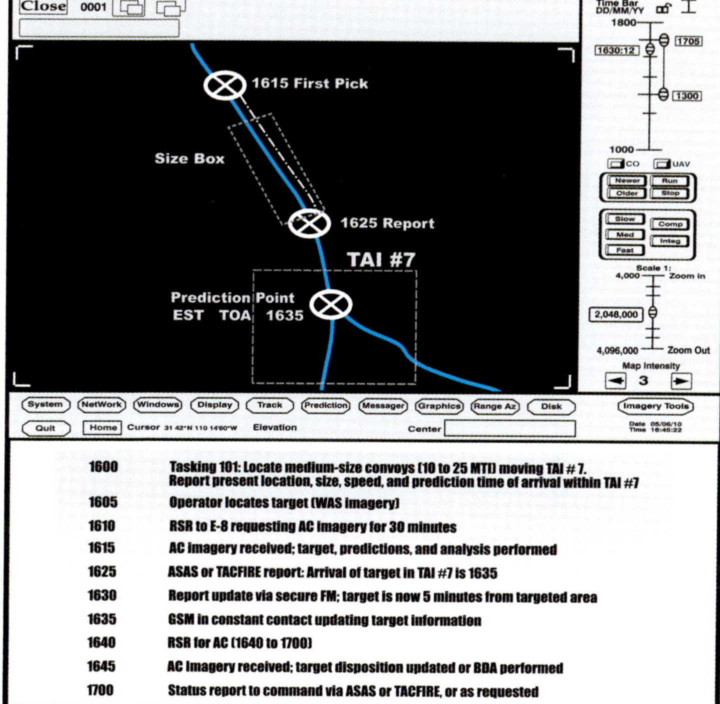

Sequence of radar screens as seen by the operators aboard the E-8.
Anderson Subtil

tracking, maritime search, and track and dismounted target MTI, the last mode designed to detect moving people.

In order to obtain better resolution, the AN/APY-7 is supplemented by the Goodrich MS-177 multispectral camera (electro-optic and infrared), installed in the rear section of the radar compartment, which is illustrated in the next chapter.

Regarding communications, the E-8 has a five-channel intercommunications system, which links all workstations and flight crew positions and provides the internal communications for the E-8. There are also three plug-in communications boxes not associated with workstations, which can be used while walking around or standing behind a workstation.

The external communications provide voice and data transmissions for line-of-sight (LOS) and non-line-of-sight (NLOS) communications.

The communications suite also has Joint Tactical Information Distribution System (JTIDS), Surveillance and Control Data Link (SCDL), and Constant Source links. JTIDS provides digitally processed intelligence reports and targeting updates into the E-8. It also allows E-8 operators to pass information to other Air Force platforms and C4I nodes such as AWACS. As it is the case with the SCDL, the JTIDS can also provide target nominations, engagement points, activity indicators, and free text messages. The JSTARS moving-target indicator (MTI) and synthetic aperture radar (SAR) data are transmitted over the SCDL to all ground station modules (GSM) within LOS.

The SCDL allows for reprocessed radar data, which includes tracked versus unknown indications, target locations, E-8 location and speed, and information time tags. In addition to these functions, it is also designed to process MTI, SAR/FTI, radar and PME status, jammer strobes and azimuth, target nominations, engagement points, activity indicators, free text messages from the GSM to the operators on the JSTARS, radar interrupt priority, reference points, and time of imagery, while the JTIDS can provide the E-8 position, air tracks, platform status, intelligence, RSRs, and ground tracks.

For its turn, the Constant Source was planned to provide graphic display of multisource, secret-level, electronic intelligence data. It processes in almost real time the threat information used by combat units/crews to plan and execute missions, and allows the soldier to access critical data provided by national and tactical intelligence sources.

Regarding the JSTARS Operation and Control Console capabilities, the subsystem allows the mission crew to access the radar data in real time on their consoles. These are basically the same capabilities that exist in the GSMs. The workstations allow operators to tailor the radar products to their needs. They can access databases to provide amplifying information on friendly and enemy order of battle (OB), receive and send free text messages over SCDL, and conduct crew coordination. The operators can perform history playback, construct SAR mosaics, track targets, and perform target position predictions. There is also a printer capability to provide prints of time-designated screen displays with complete annotations. The mission crew can zoom in or out on the surveillance area by adjusting the scale parameters on the cartographic image displayed on their console screens. The operator can measure and display distance and azimuth between specific geographic points contained in the database and between selected targets. All necessary graphics, including standard military symbols, can be drawn or retrieved from the database for display on the screens. The symbology is tailored to the specific requirements of a given mission and can include an extensive list of items, such as major elements of friendly forces; fire support coordination measures such as troops in contact; forward edge of the battle area; forward lines of one's own troops; fire support coordination lines; restricted operating zones; combat air patrols; landing/drop zones; area of operations; no-fire zones, free fire zones, and restricted fire zones; applicable airspace control measures such as air refueling tracks and corridors; location of friendly airfields; designated areas for recovery or search and recovery; special-operations-forces locations; current day's named areas of interest and orbit areas of operations; downed aircrew locations; major elements of enemy forces; location of enemy airfields; air order of battle; combat air patrol activity; Integrated Air Defense System elements; defensive missile order of battle; electronic order of battle, including early-warning and ground control intercept sites and visual observer locations; air defense artillery; surface-to-air firings and sources; ground order of battle; surface-to-surface missile locations and launchers; naval order of battle; asymmetric events (e.g., directed-energy weapon incidents, improvised explosive devices, minefields); chemical-, biological-, radiological-, nuclear-, and explosives-contaminated areas; a legend depicting all symbols and associated captions; classification and control markings; current-as-of-date time groups; and, finally, supplementing and updating the situation displays as intelligence becomes available.

The operators can also conduct time and route prediction. The estimated time of target arrival to selected points or along selected routes can be predicted along with the route predictions.

The consoles also have specific functions as time compression function; that is, the ability to record MTI radar data over time and then fast-forward the data frames, which is useful in tracking the target start point, route, and end point over a selected period, and also in providing the cue on where to image the target with SAR, and time integration function, which is the ability of the system to overlay successive frames on top of each other over a selected period and display them all at one time on the screen. This allows for the rapid identification of main supply routes. This function, present in all GSMs, also provides cross-cuing for other sensors (or SAR) to confirm and target the locations.

Self-Defense Suite (SDS) onboard the E-8 is designed to provide some measures of defense against air and ground threats, providing the aircrew with continuous situational awareness and having an end-game countermeasures package designed

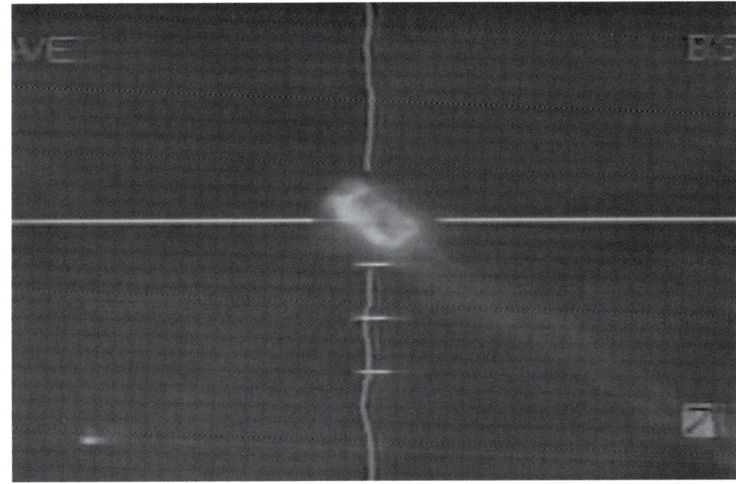

A target car as seen by the screen of an operator aboard the JSTARS. *Editing work by Anderson Subtil from a Northrop Grumman promotional video*

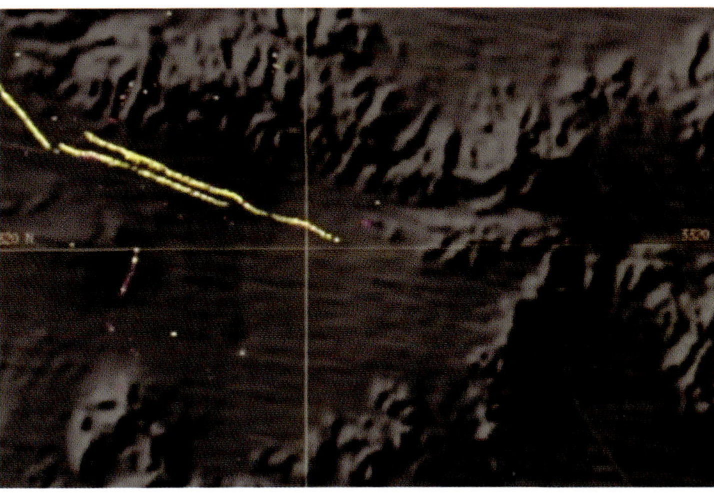

An enemy convoy as seen by the screen of an operator aboard the JSTARS. *Editing work by Anderson Subtil from a Northrop Grumman promotional video*

to protect the aircraft from various types of air- and ground-launched missiles. It receives threat information over JTIDS and Constant Source terminals from AWACS; Airborne Command and Control Center (ABCCC); RIVET JOINT; other air- and ground-based sensors; and command, control, communication, and intelligence nodes. The navigator staffs this position and conducts all navigation functions in addition to the SDS functions.

Once obtained, all the imagery obtained by the JSTARS mission equipment is relayed to the ground segment of the system. Throughout the E-8's evolutionary path, this segment has been successively named Interim Ground Station Module (IGSM), Light Ground Station Module (LGSM), and Medium Ground Station Module (MGSM). However, these have been replaced by the TSQ-179 (V) Common Ground Station (CGS), of which the US Marine Corps acquired two examples in 1997.

The CGS derives from the LGSM, AN/TSQ-178, and radar data from the JSTARS aircraft are broadcast to multiple CGS via a secure SCDL. In addition to receiving moving-target indicator (MTI) and synthetic aperture radar (SAR) data, the baseline CGS receives Hunter UAV video, intelligence broadcasts, and national imagery and disseminates information to Army intelligence and fire support nodes. Block 10 CGS provides UHF SATCOM connectivity to the JSTARS aircraft, while Block 20 CGS integrates the CGS with the local-area network of the Army Battle Command System Tactical Operations Center to improve and expand the dissemination of CGS products.

A typical mission requires the CGS crew have a minimum amount of preplanned mission and baseline taskings to support the intelligence collection process. These taskings are given to the team leader during mission briefs. The CGS receives immediate taskings and mission changes from the Army G2/G3, collection manager, or Army S2/S3 (or the FSE), depending on the echelon and unit being supported.

The CGS operators work these preplanned and baseline taskings during the mission execution. Their products are the reports and imagery copies that go to their supported unit. If there are no immediate taskings received, the operators continue to work baseline taskings. If an immediate tasking is received, the CGS operators complete the baseline tasking and then start working on the immediate tasking. Once the operator completes the immediate tasking, he works baseline taskings until another immediate tasking is received. The collection manager must provide this information one to two hours prior to the start of CGS operations.

However, this system will be replaced by the Distributed Common Ground System–Army (DCGS-A). According to various sources, the new system enables unprecedented timely, relevant, and accurate targetable data to the warfighter. DCGS-A supports the Army's Unified Mission Command System and provides access to information and intelligence to support battlefield visualization and ISR management in accordance with the Army Common Operating Environment. It provides information discovery,

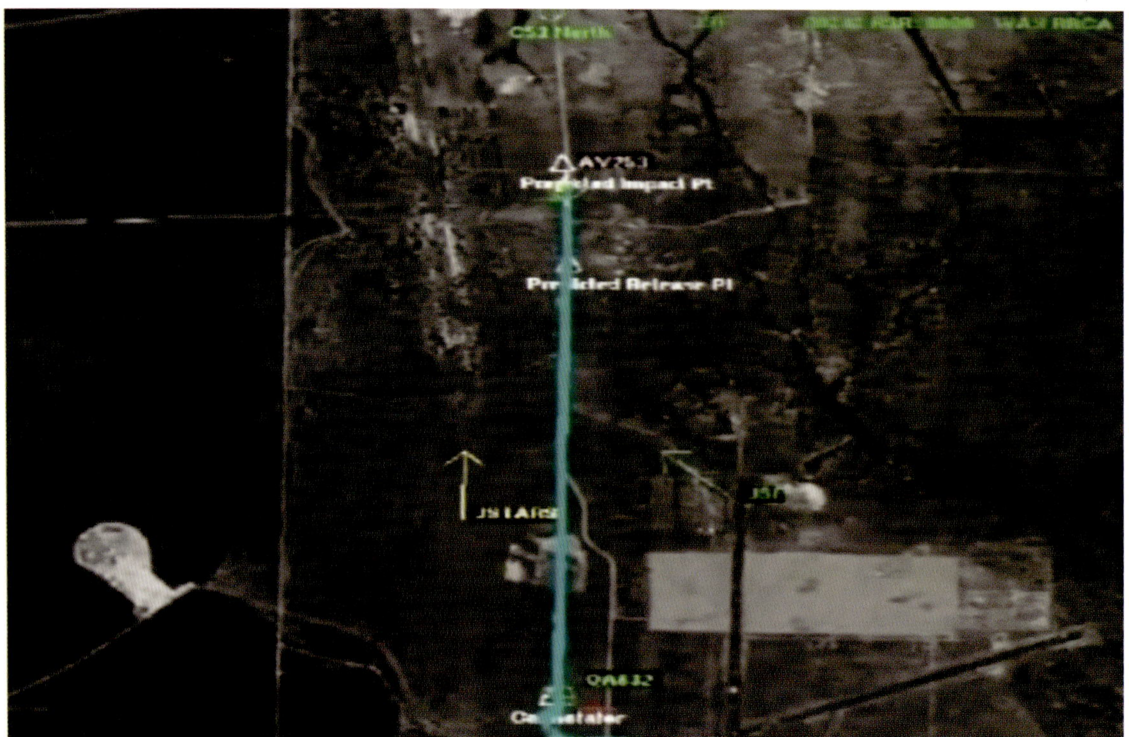

Enemy forces as they appear in the screen of an operator aboard the E-8 JSTARS. *Editing work by Anderson Subtil from a Northrop Grumman promotional video*

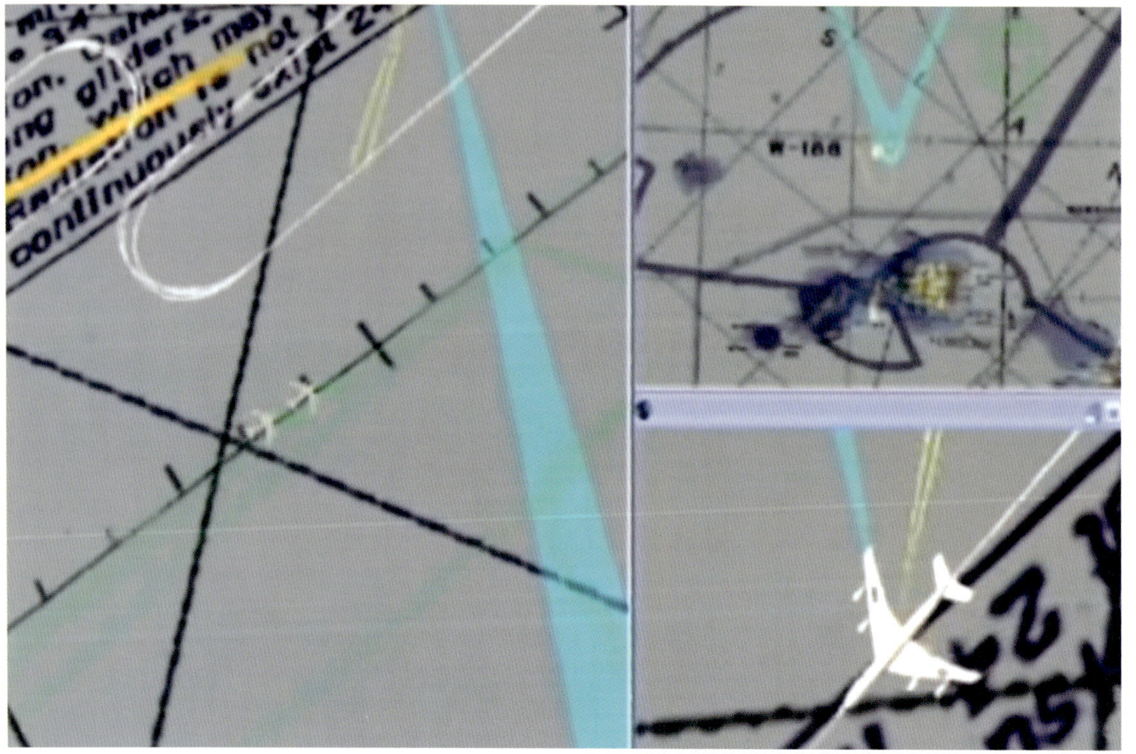

Samples of typical orbits flown by the E-8 during a surveillance mission. *Editing work by Anderson Subtil from a Northrop Grumman promotional video*

The E-8A, a GSM truck, and its development crew. *Martin Kleiner*

collaboration, production, and dissemination to commanders and staffs globally.

DCGS-A assumes life cycle management responsibility and consolidates or replaces the operational capabilities provided by several post–Milestone C programs of record and fielded quick-reaction capabilities. The Army fields DCGS-A capability on various hardware (HW) platforms using a consolidated software baseline. HW platforms range from single laptops to multiserver, transportable configurations able to process and store the enormous volumes of data that DCGS-A must manage. DCGS-A's modular, open-systems architecture allows rapid adaptation to changing circumstances and the ability to have and collect intelligence while on the move. It is said that more than 700 information sources can be accessed by this new system, which has three configurations: (1) Fixed, which leverages the power and stability of sanctuary for the most-complex processing and analytic tasks. Additionally, it provides the greatest historical data repository. (2) Mobile, which provides a tactical, deployable capability to deliver responsive, forward support to commanders from BN through operational headquarters. Analytical tools, sensor inject, data storage, and integration with other BCS are the highlights of the Mobile configuration. (3) Embedded, which enables the connection of the intelligence enterprise with the battle command network.

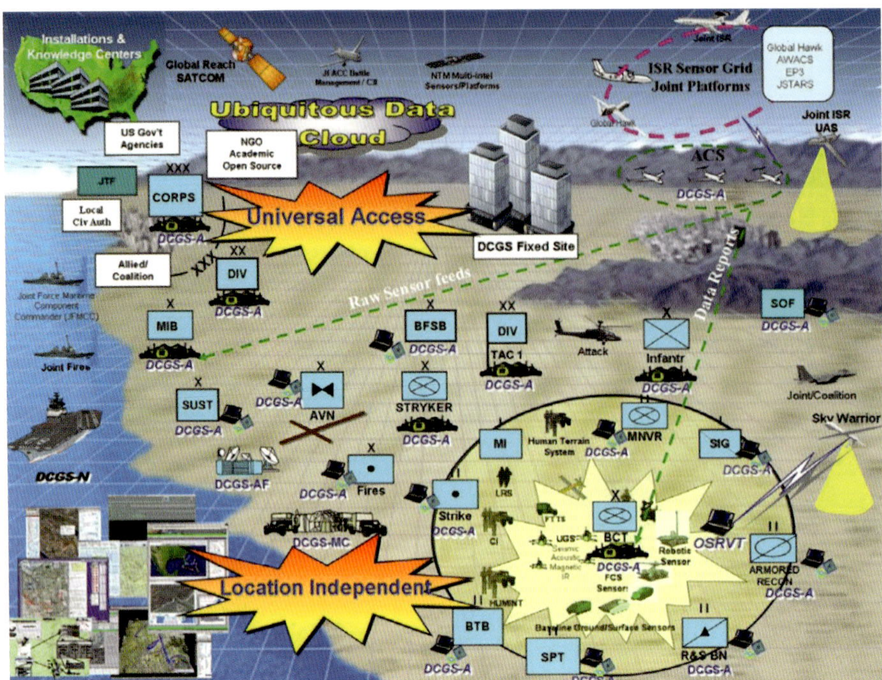

Figure 1-1. DCGS-A OV-1 and Future Force.

The capabilities of the Distributed Common Ground System–Army. *Army Distributed Common Ground System (DCGS A) Commander's Handbook*

Because the E-8Cs are commanded by the Air Combat Command (ACC), this authority designates the corps commanders or units to be supported, who in turn will determine the required coverage area, effective employment times of coverage, and radar priorities. When required, the ACC deploys with adequate aircraft to provide twenty-four-hour coverage of the designated GRCA.

E-8 subsystems' specific parts and personnel replacements would be pushed into the theater supply and replacement system from the supporting wing to the operations squadron. Examples

The Distributed Common Ground System–Army, successfully shared full-motion video between Air Force assets, directly to the Army's DCGS-A Tactical Ground Station. This capability enables joint intelligence sharing across the battle space as part of the Enterprise Challenge 13 exercise. *Sgt. 1st Class Kristine Smedley*

US Air Force Team JSTARS aircraft maintenance personnel perform a foreign-object-damage (FOD) walk on the flight line while an E-8C Joint STARS sits in the background at Robins Air Force Base, Georgia, on July 20, 2017. *US Air National Guard photo by SMSgt. Roger Parsons*

USAF 2nd Lt. Amanda Villegas, *left*, and 1st Lt. Aryk Bingham-Hill, air weapons officers with the 330th Combat Training Squadron, perform a preflight inspection at their operator work stations aboard an E-8C during Exercise Razor Blade 18-02, on February 8, 2018. *US ANG photo by SMSgt. Roger Parsons*

Usual briefing prior to a mission. *US Air National Guard photo by SMSgt. Roger Parsons*

USAF SSgt. Justin Estergard, a crew chief with the 461st Aircraft Maintenance Squadron, inspects an engine inlet on an E-8C Joint STARS while doing a prelaunch inspection during Exercise Razor Blade 18-02, planned to assess the unit's readiness. *US Air National Guard photo by SMSgt. Roger Parsons*

Air-conditioning equipment is positioned close to an E-8C prior to a mission. *US Air National Guard photo by SMSgt. Roger Parsons*

USAF Senior Airman Trevor Holeman, a communications and navigation technician with the 116th Aircraft Maintenance Squadron, Georgia ANG, performs maintenance on a weather radar display unit on an E-8C JSTARS prior to a launch during Exercise Razor Blade 18-02. *US Air National Guard photo by SMSgt. Roger Parsons*

US airmen from Team JSTARS pick up their mobility bags in preparation for a simulated deployment during Exercise Razor Blade 18-02 at Robins Air Force Base, Georgia, on February 6, 2018. *US Air National Guard photo by SMSgt. Roger Parsons*

US airmen from Team JSTARS in process through the 78th Air Base Wing Installation Personnel Readiness center for a simulated deployment during Exercise Razor Blade 18-02. *US Air National Guard photo by SMSgt. Roger Parsons*

USAF SSgt. Julian Dash, with the 461st Air Control Wing (ACW), loads E-8C Joint STARS aircraft tires on a trailer by using a forklift for a simulated deployment during Exercise Razor Blade 18-02. *US Air National Guard photo by SMSgt. Roger Parsons*

of support would be radome and SCDL equipment and software upgrades support. The wing would also be responsible for locating any continental United States (CONUS) parts that may be required.

These requirements are passed to the Air and Space Operations Center (ASOC) through the Army BCE (Army Battle Coordination Element) for Joint STARS mission planning and tasking. The joint force air component commander (JFACC) determines the number of JSTARS aircraft and orbits needed to provide the required coverage. In support of the joint force commander's guidance and campaign objectives, the joint force land component commander (JFLCC) allocates the corps commanders or units to be supported, and for how long.

It should be noted that JSTARS has no inherent identification capability, so all target identification requires off-board cross-cueing. Although JSTARS does not provide airspace-monitoring positive radar control of aircraft, it can provide procedural control of aircraft by using time, altitude, geographic (lateral) separation, and data links for aircraft deconfliction. The deconfliction method may be delineated in real-time tactical briefs or specific

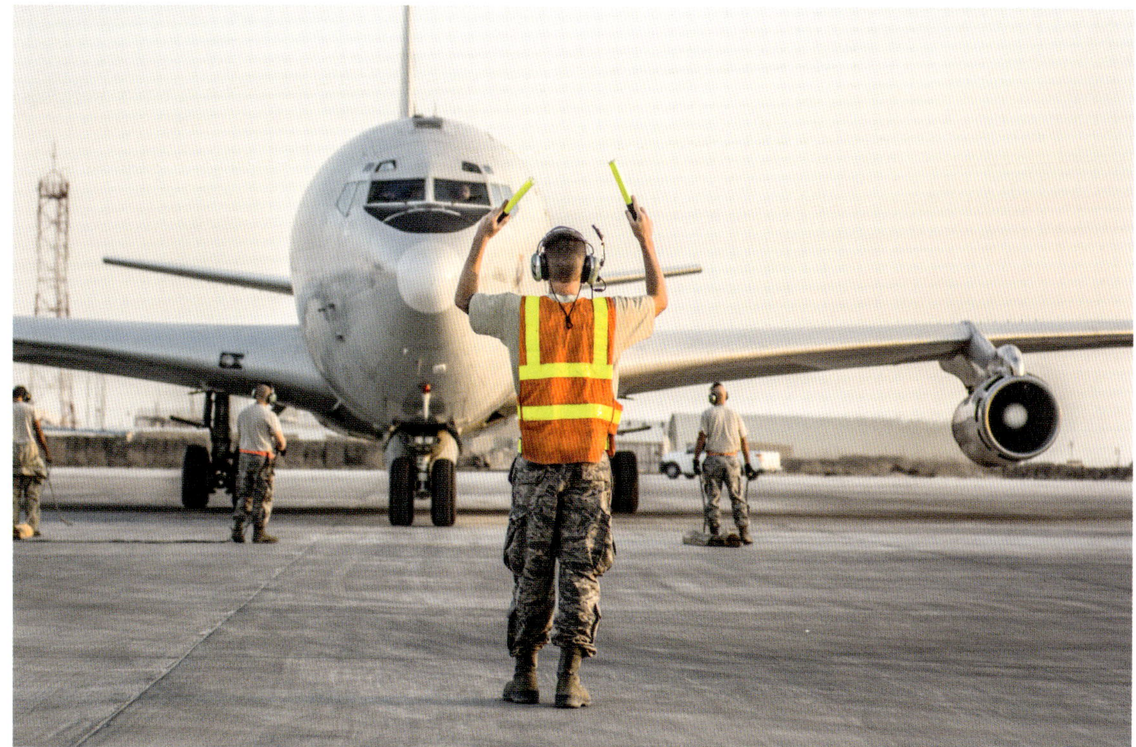

US Air Force Senior Airman William Connelly, assigned to the 7th Aircraft Maintenance Unit, marshals an E-8C Joint Surveillance Target Attack Radar System after a mission at Al Udeid Air Base, Qatar, on November 6, 2017. *US Air National Guard photo by SSgt. Patrick Evenson*

Airman 1st Class Jeremy Cole, 379th Expeditionary Aircraft Maintenance Squadron crew chief, marshals an E-8C Joint Surveillance Target Attack Radar System that landed October 20, 2016, at Al Udeid Air Base, Qatar, following a mission supporting Operation Inherent Resolve. *US Air Force photo/Senior Airman Miles Wilson/Released*

Senior Airman Miles Wilson marshals an E-8C Joint Surveillance Target Attack Radar System that landed on October 20, 2016, at Al Udeid Air Base, Qatar, following a mission supporting Operation Inherent Resolve. *US Air Force*

instructions. When operating in the proximity of other aircraft, these must not penetrate within 1,000 feet vertically or 1,500 feet horizontally of the JSTARS.

Most operational tactics are classified, but it is well known that during the early years of operations, a JSTARS crew typically used to establish an orbit that was distant from the requested search area by approximately 26.9 nautical miles. However, as doctrinal aspects evolved from the real operations, radar surveillance now is distributed on up to fourteen individual areas, each measuring up to 4 square miles, which allows radar cover on multiple areas for a considerable amount of time.

From the crew perspective, a typical mission begins with briefings. A first practical step toward an operational flight is the search for any kind of objects that may cause "foreign object damage." This search is performed by ground crews, who walk carefully throughout the full extension of aprons, taxiways, and runways. Another phase for a successful deployment is the correct transport of all necessary items. This was simulated during

Part of an E-8C aircrew commences to turn its mission equipment off after having arrived back to Al Udeid Air Base following an operational flight. *US Air Force*

Aircrew descends from its aircraft following another operational sortie. *Senior Airman Miles Wilson / US Air Force*

SSgt. Richard Routh, 116th Aircraft Maintenance Unit, 379th Expeditionary Aircraft Maintenance Squadron, works on an E-8C JSTARS. After the aircraft flies either 700 hours or for nine months, whichever comes first, the aircraft requires three days of preventive maintenance, otherwise known as a 700-hour inspection. *US Air Force*

Members of the 379th Expeditionary Aircraft Maintenance Squadron conduct a post–flight systems check on an E-8C Joint Surveillance Target Attack Radar System on October 20, 2016, at Al Udeid Air Base, Qatar, following a mission supporting Operation Inherent Resolve. *US Air Force photo / Senior Airman Miles Wilson*

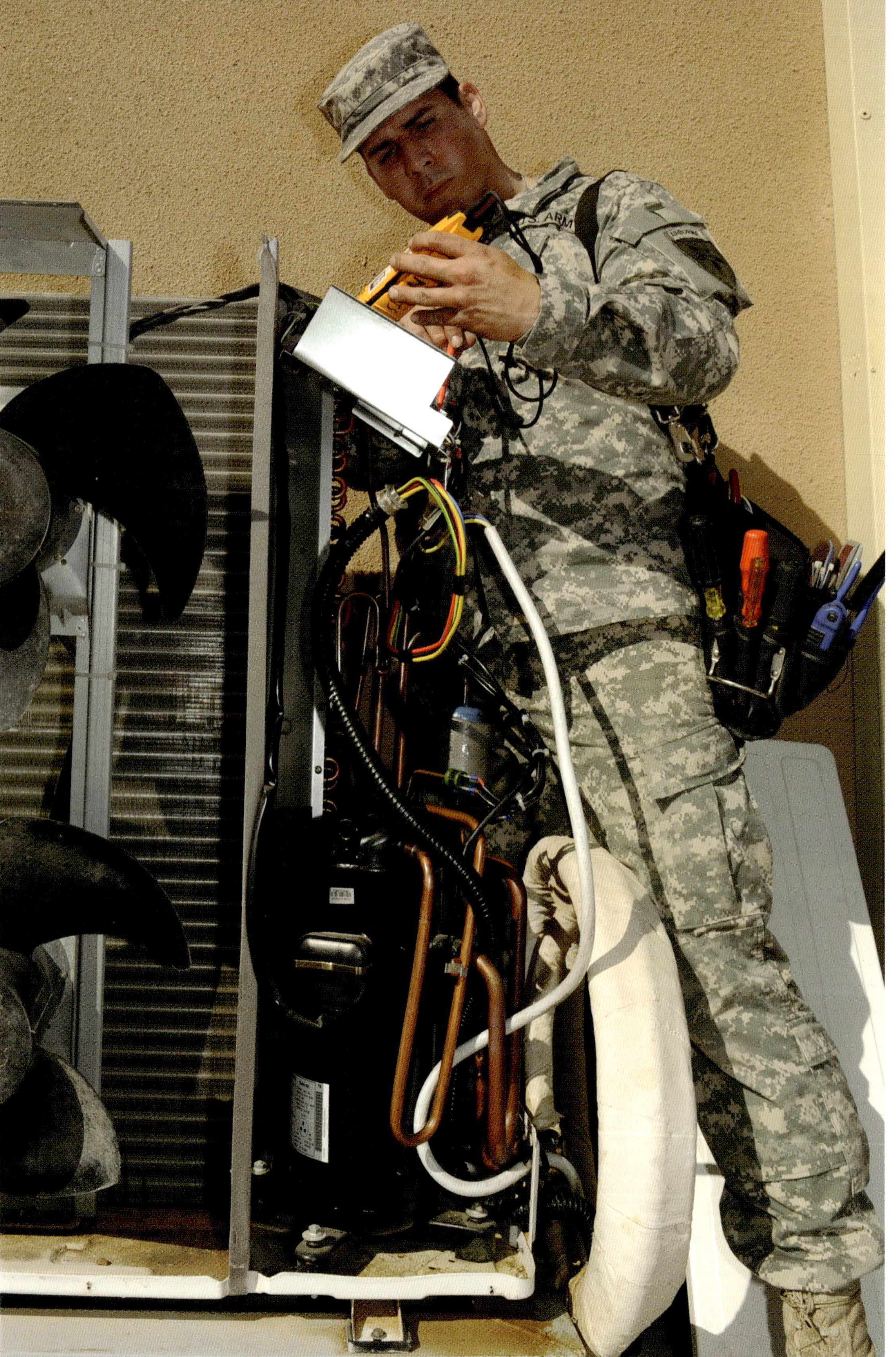

Army Spc. Jorge Diaz, US Central Command heating, ventilation, and air-conditioning technician, checks an AC unit's power at an undisclosed Southwest Asia location, on March 17, 2010. *US Air Force photo by TSgt. Michelle Larche*

US Air Force SSgt. Daniel Spear, aerospace propulsion craftsman, 7th Aircraft Maintenance Unit, performs postflight inspections of the intake and exhaust systems on an E-8C JSTARS after a mission at Al Udeid Air Base, Qatar, on November 6, 2017. *US Air National Guard photo by SSgt. Patrick Evenson*

SSgt. Michael Johnson, 116th Aircraft Maintenance Unit, 379th Expeditionary Aircraft Maintenance Squadron, changes a filter in the generator as a part of the inspection. *US Air Force*

SSgt. Juan Hernandez (*right*) and Senior Airman Nick Bentler, 379th Expeditionary Maintenance Squadron aerospace maintenance, build up and overhaul an E-8C JSTARS aircraft wheel-and-tire assembly at an undisclosed Southwest Asia location, on March 16, 2010. *US Air Force photo by TSgt. Michelle Larche*

SSgt. Juan Hernandez (*right*) and Senior Airman Nick Bentler, 379th Expeditionary Maintenance Squadron aerospace maintenance, build up and overhaul an E-8C JSTARS aircraft wheel-and-tire assembly at a nondisclosed Southwest Asia location, on March 16, 2010. *US Air Force photo by TSgt. Michelle Larche*

US Air Force TSgt. Bradley Heller, crew chief assigned to the 7th Aircraft Maintenance Unit, performs through-flight inspections on an E-8C Joint Surveillance Target Attack Radar System after a mission at Al Udeid Air Base, Qatar, on November 6, 2017. *US Air National Guard photo by SSgt. Patrick Evenson*

Exercise Razor Blade 2018, which assessed the unit's readiness status for an operational deployment.

Once an E-8C returns to land after an operational mission, it is submitted to an extensive check to make it ready for the next flight. This is carried out by the flight and ground crews and consists, among other procedures, of turning the mission equipment off and on, and an external inspection looking for any flaws that could prevent the aircraft from performing its mission.Finally, in order to provide a better comprehension of the real-world operations involving the E-8C JSTARS, the following images were taken from two videos: one was shot at Al Udeid Air Base in Qatar in 2015, while the remaining one was shot two years later during Exercise Northern Strike 17, held at Alpena Combat Readiness Center, Michigan, in August 2017.

An E-8C being prepared to depart from Al Udeid Air Base for a night sortie in 2015. *Editing work by Anderson Subtil from a US Air Force video*

A close-up view of the cockpit of the same aircraft. *Editing work by Anderson Subtil from a US Air Force video*

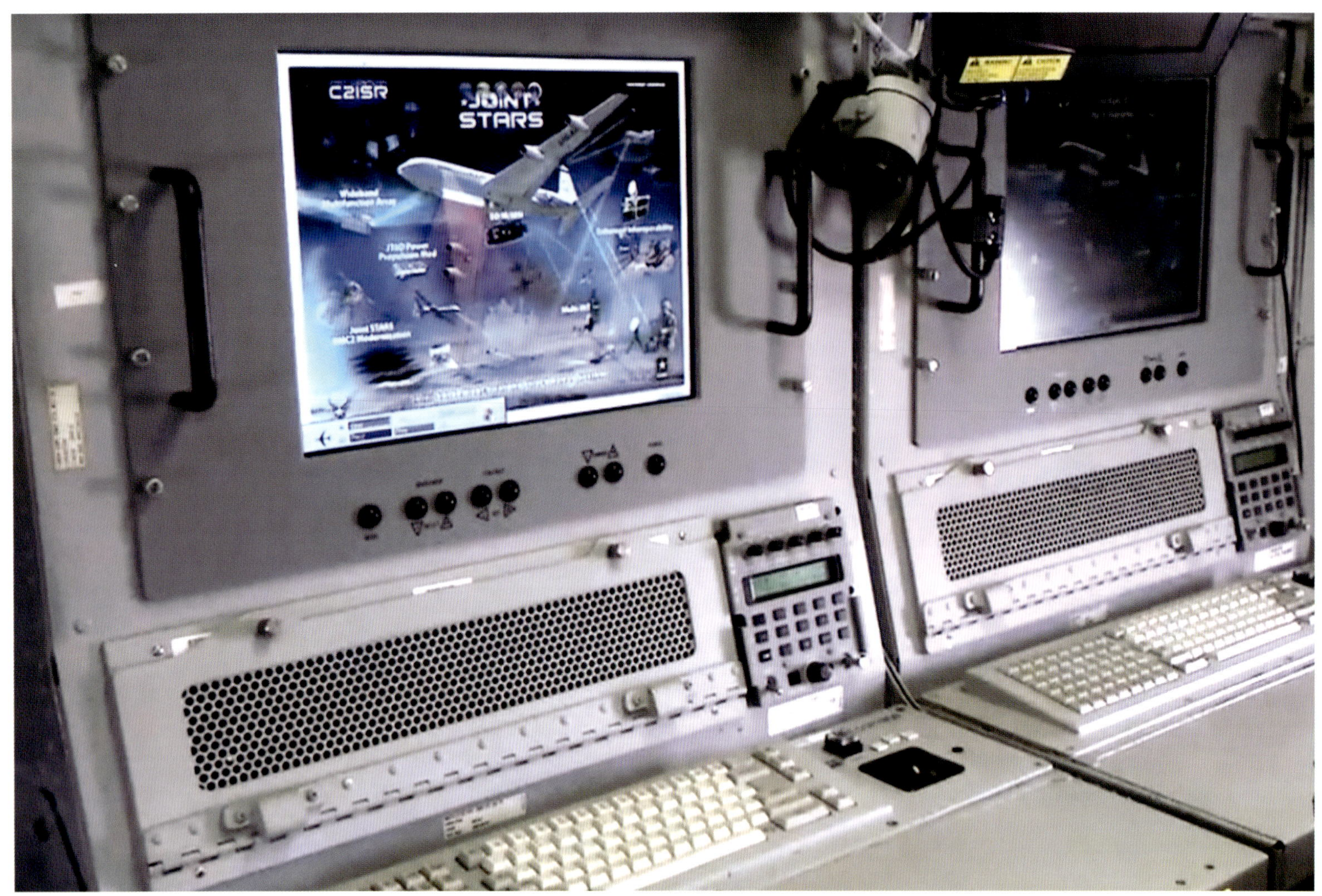

A partial view of some screens of the same aircraft. *Editing work by Anderson Subtil from a US Air Force video*

Operators getting ready for a night sortie aboard an E-8C. *Editing work by Anderson Subtil from a US Air Force video*

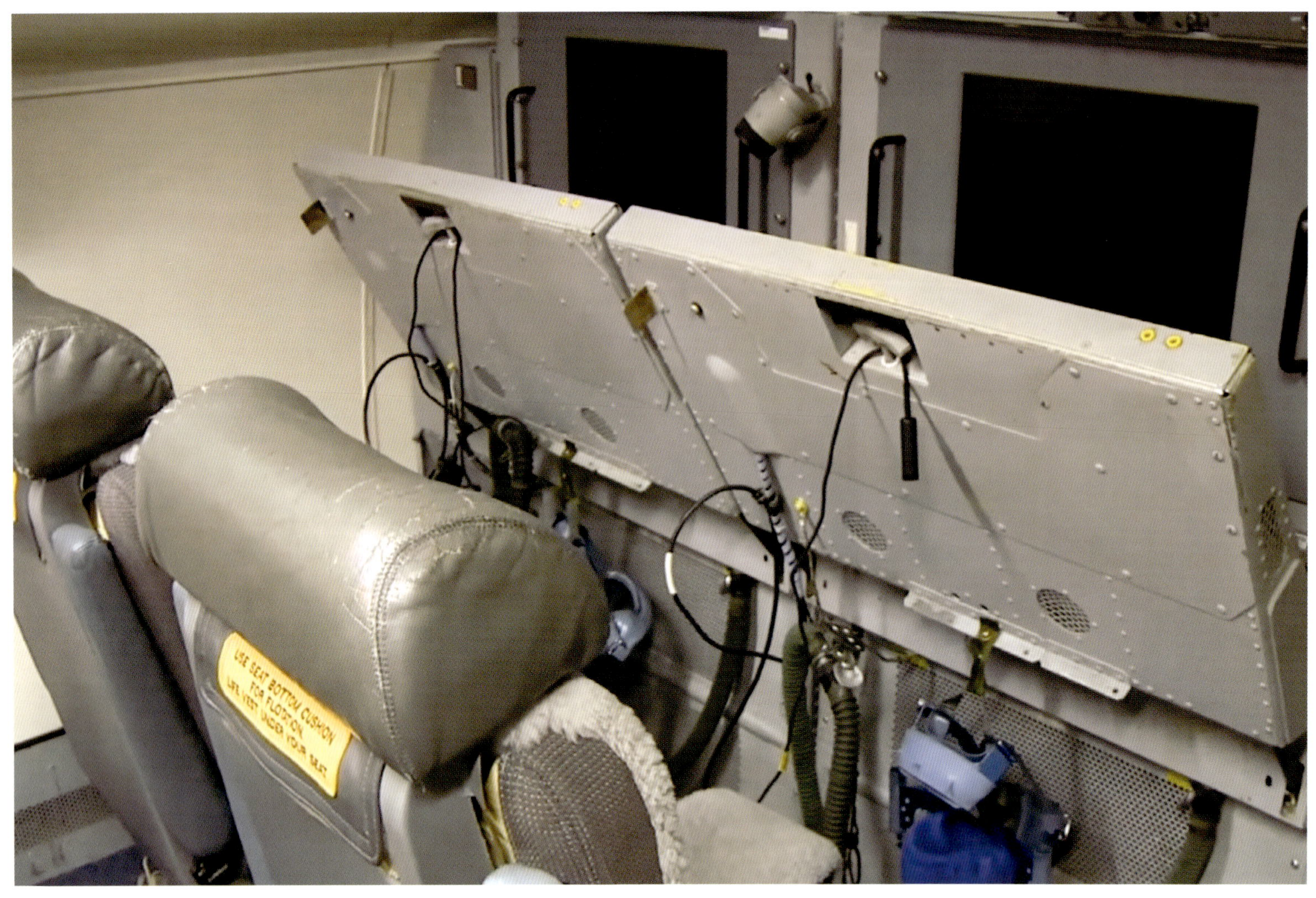

Folded tables of two workstations aboard an E-8C JSTARS aircraft. The emergency oxygen masks are noteworthy. *Editing work by Anderson Subtil from a US Air Force video*

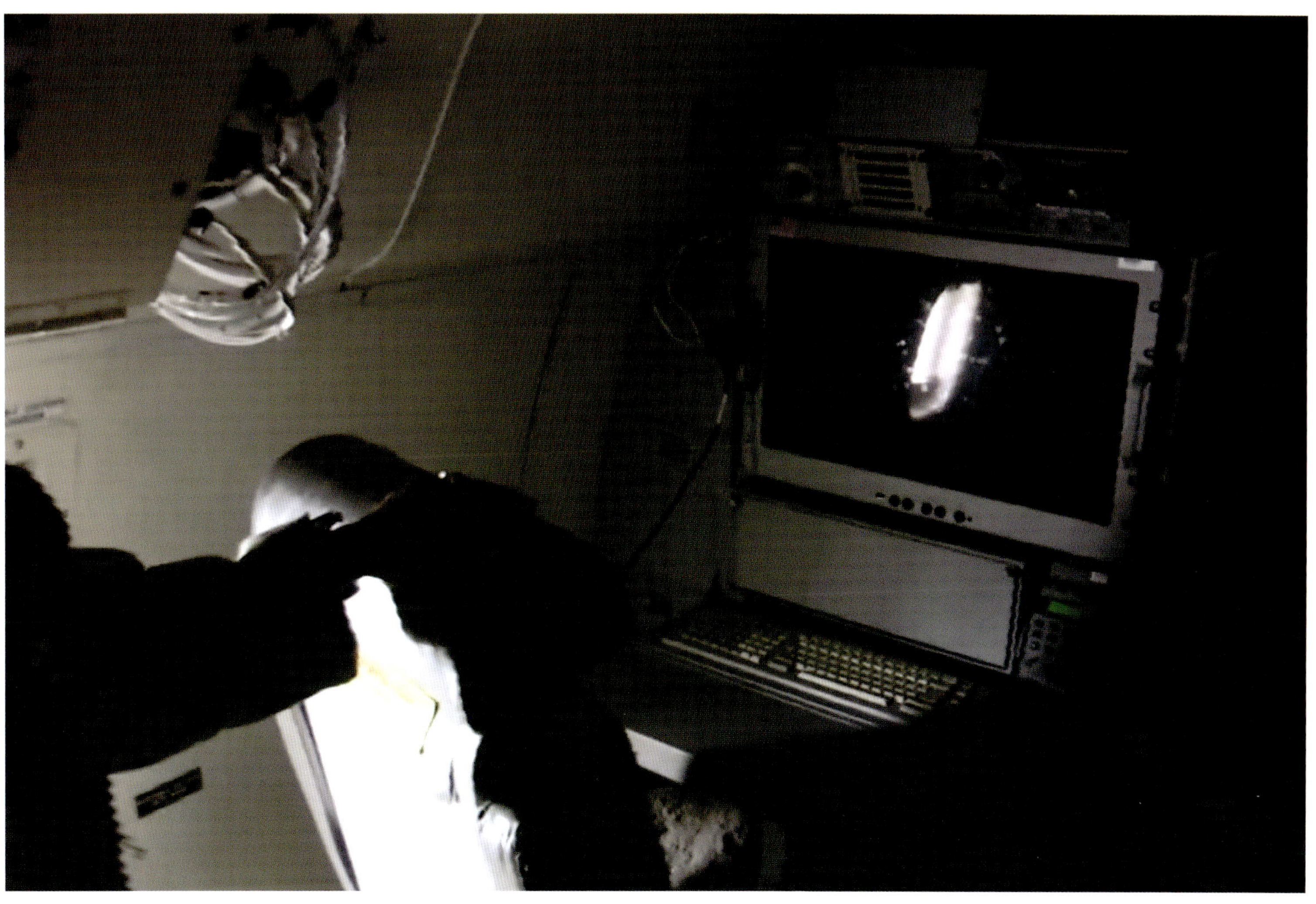

A darkened view of one of the workstations aboard an E-8C during Exercise Northern Strike 17, held at Alpena Combat Readiness Center, Michigan, in August 2017. *Editing work by Anderson Subtil from a US Air Force video*

A closer view of some workstations aboard an E-8C JSTARS during the same exercise. *Editing work by Anderson Subtil from a US Air Force video*

An E-8C mission crew component checks its workstation for Exercise Northern Strike 17. *Editing work by Anderson Subtil from a US Air Force video*

A view of the E-8C aisle toward the flight deck during Exercise Northern Strike 17. *Editing work by Anderson Subtil from a US Air Force video*

A somewhat darkened view of the same place. *Editing work by Anderson Subtil from a US Air Force video*

The crew rest seats aboard the E-8C JSTARS. *Editing work by Anderson Subtil from a US Air Force video*

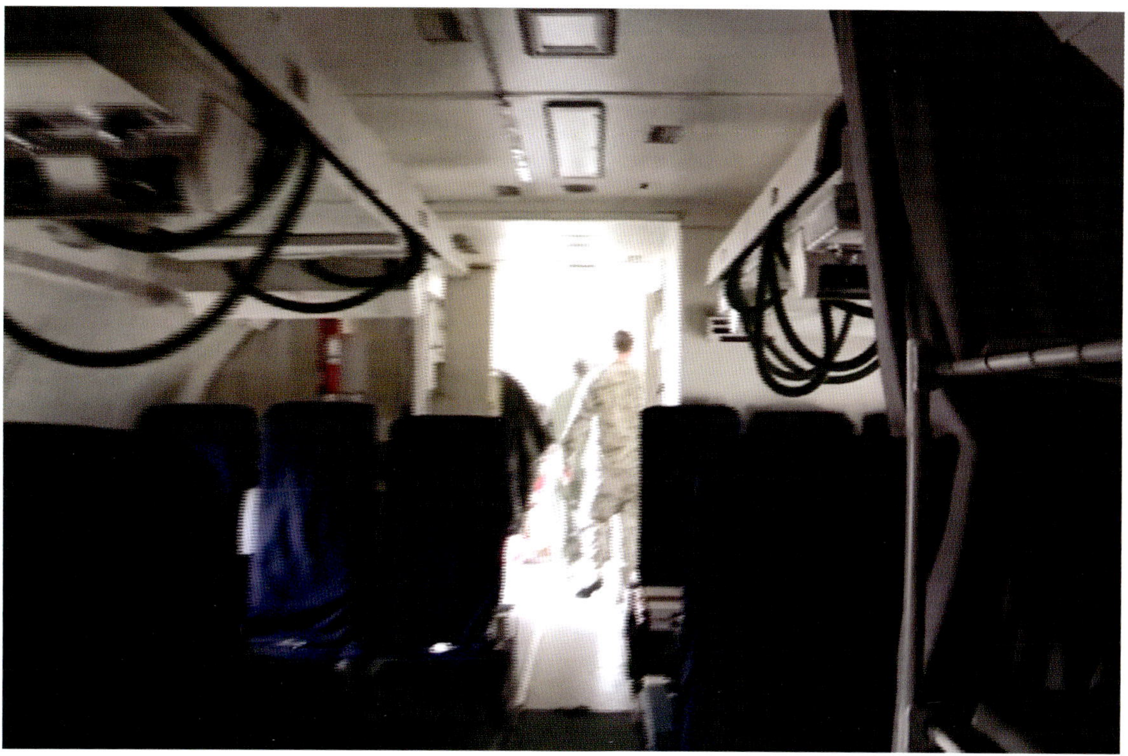

A clearer view of the frontal section of the E-8C JSTARS, with the port main entrance opened. *Editing work by Anderson Subtil from a US Air Force video*

The E-8C 97-0201 tail. *Editing work by Anderson Subtil from a US Air Force video*

A good shot of the E-8C 97-0201 frontal section. *Editing work by Anderson Subtil from a US Air Force video*

The port main entrance of the E-8C 97-0201 with an access ladder. *Editing work by Anderson Subtil from a US Air Force video*

The emblems of the Air National Guard, 116th ACW, and Air Combat Command applied over the windows. *Editing work by Anderson Subtil from a US Air Force video*

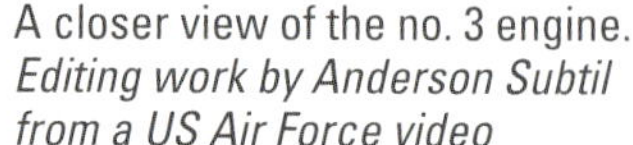

A closer view of the no. 3 engine. *Editing work by Anderson Subtil from a US Air Force video*

A distant view of the E-8C 97-0201, which took part in Exercise Northern Strike 17. *Editing work by Anderson Subtil from a US Air Force video*

A closer shot of the same aircraft. *Editing work by Anderson Subtil from a US Air Force video*

CHAPTER 3

The Latest Upgrades

Throughout its operational career, the JSTARS fleet has been subjected to several modernization programs in order to keep it ready to deal with the most-recent threats. Other than the replacement of the old Pratt & Whitney TF33-102Cs with the Pratt & Whitney JT8D-219s, which failed in 2011 due to the lack of funds, these programs commenced in 2005, when all the E-8Cs were converted to the Block 20 and the JSTARS fleet was entered in a program to increase its overall performance.

In 2010, a contract to incorporate the Multi-Platform Radar Technology Insertion Program was inked. This improvement comprises an antenna, the radio frequency electronics, and the signal processor and provides, among other improved operational gains, enhanced wide-area surveillance (WAS) system capabilities to the warfighter. It was followed in 2011 by the introduction of the MS-177 multispectral camera (planned to deliver high-resolution imagery) and the Multifunctional Information Distribution System, Joint Tactical Radio System (MIDS-JTRS), which improves the Link-16 connectivity with other communication protocols.

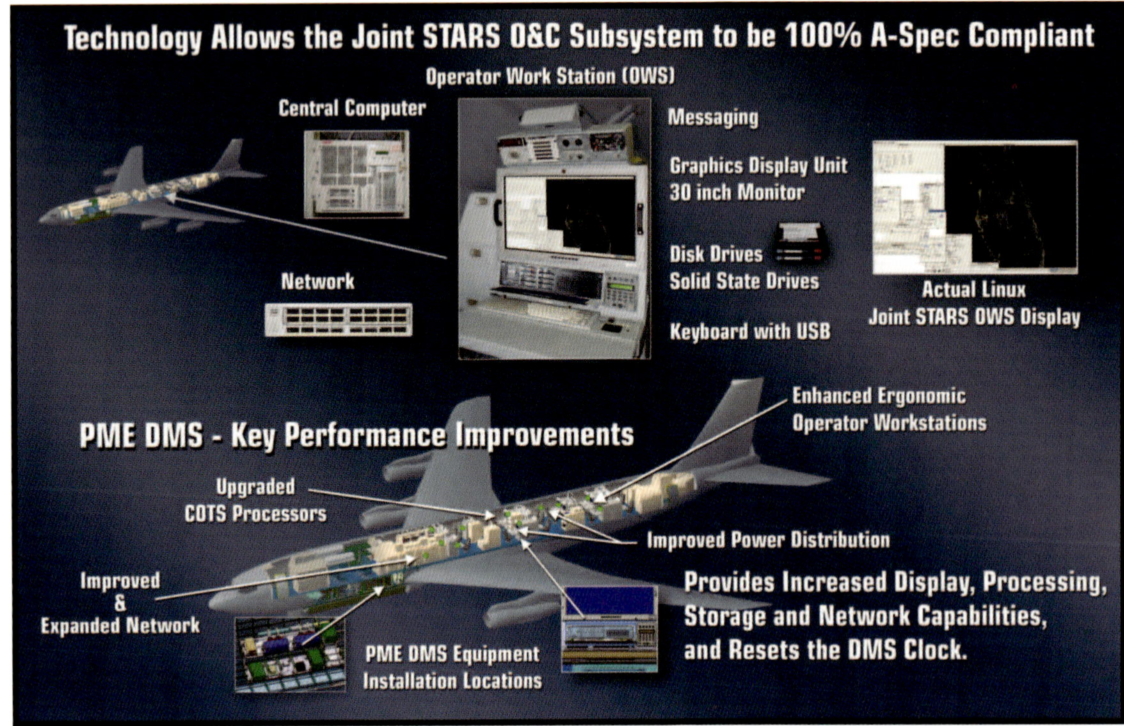

The new workstations planned to be installed aboard the E-8. *Northrop Grumman*

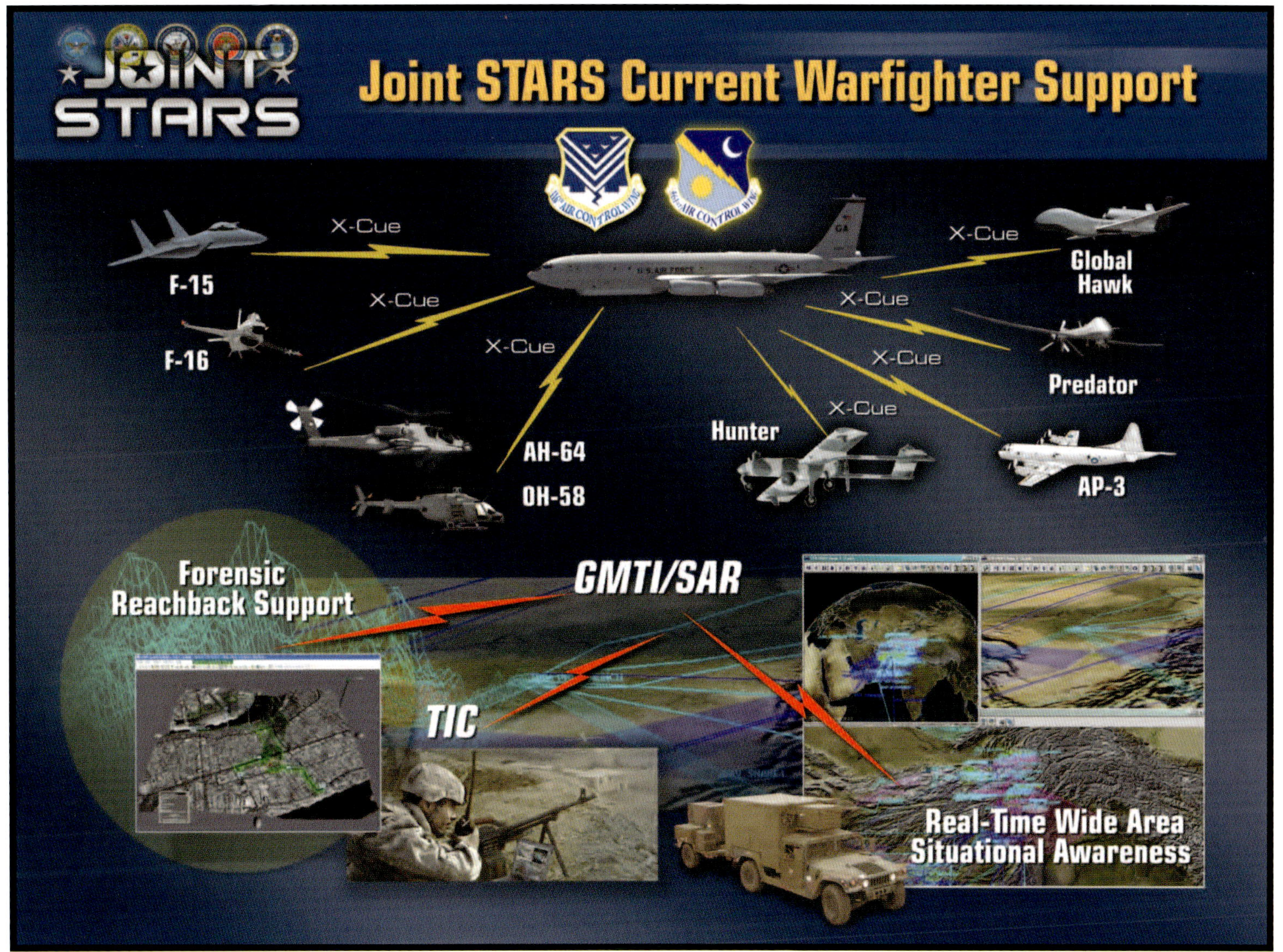

The connection capabilities scheduled to be incorporated into the JSTARS. *Northrop Grumman*

The capabilities of the MS-177 Multispectral Camera. *Northrop Grumman*

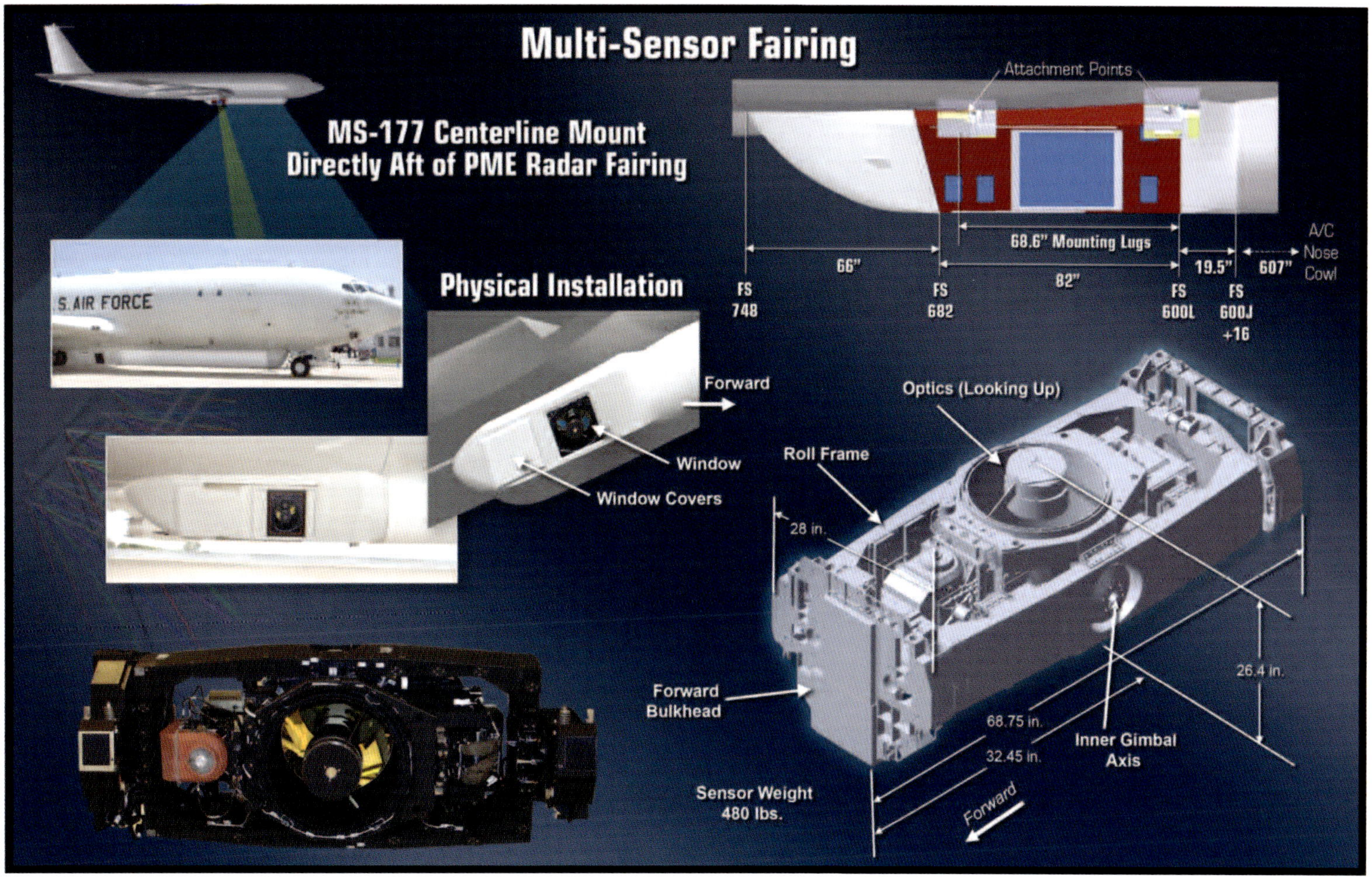

Detailed view of the MS-177 Multispectral Camera. *Northrop Grumman*

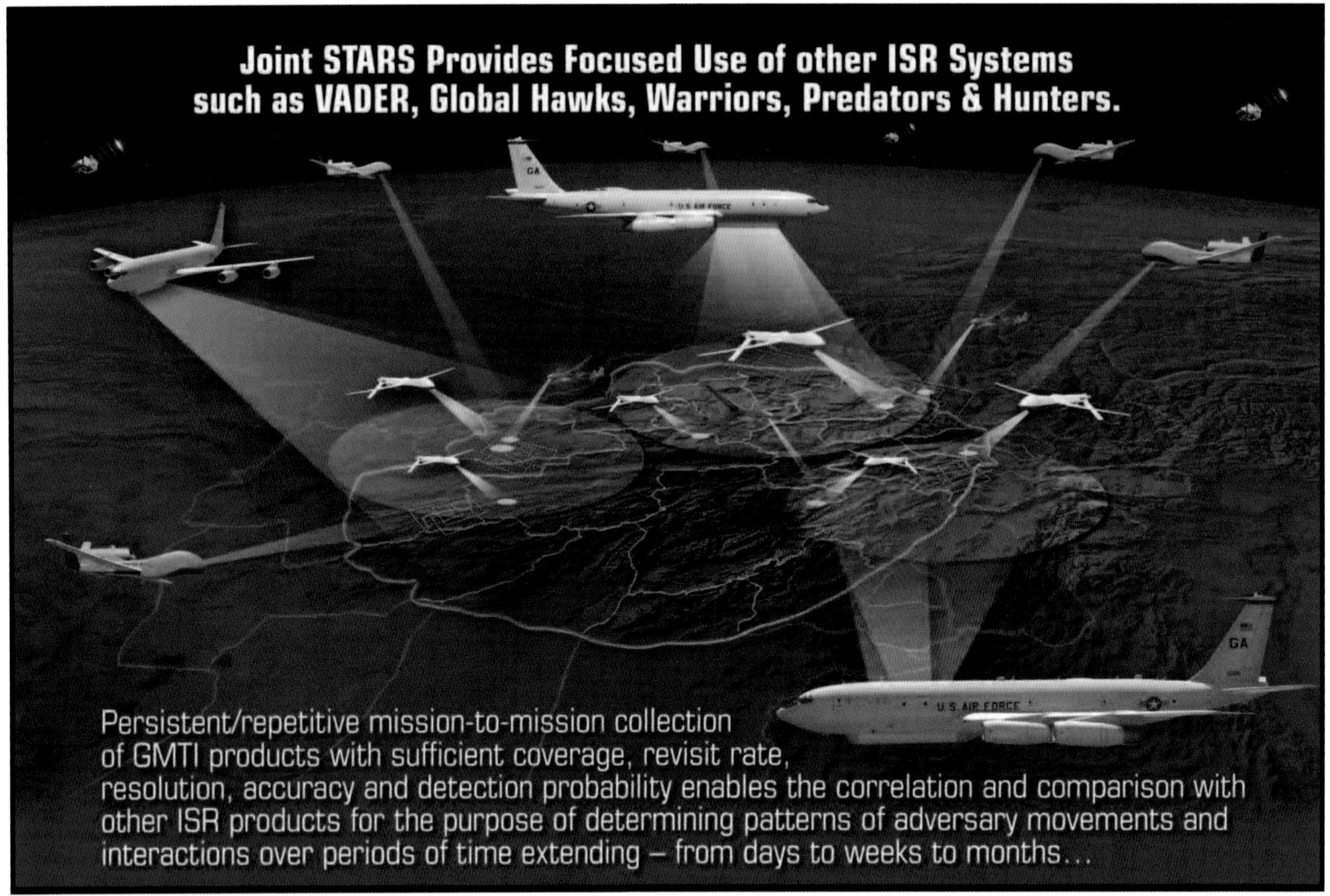

Graphic demonstration of the E-8C's ability to share data with other ISR assets. *Northrop Grumman*

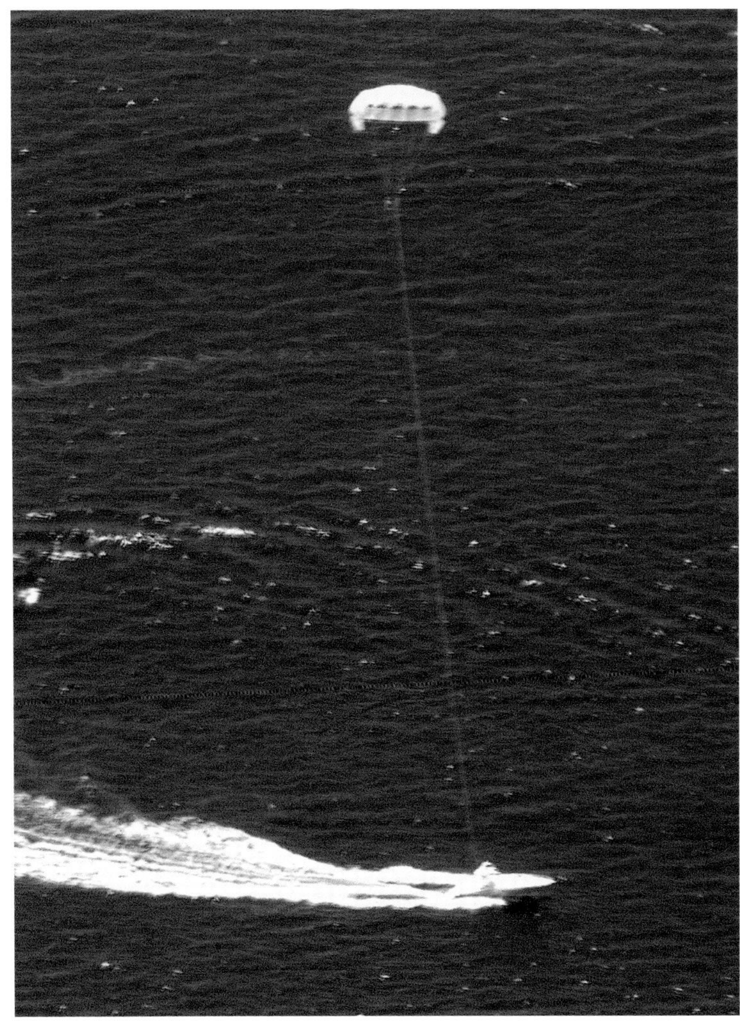

Image obtained by the MS-177 Multispectral Camera. *Northrop Grumman*

Another type of image obtained by MS-177. *Northrop Grumman*

A thermal image captured by the MS-177 Multispectral Camera. *Northrop Grumman*

In 2013, it was decided to equip the JSTARS with new workstations and radar signal processor computers, in a program yet to be terminated in 2020. The E-8C radar modernization was concluded and the SATCOM was added in that same year.

In 2014, the Multi Agency Communication Capability was added and the SATCOM was upgraded, while another program, "Prime Mission Equipment Diminishing Manufacturing Sources," which was planned to replace some items of the navigation and communication suites that had become hard to find, started to be implemented.

Among these items were communications, navigation, surveillance, and air traffic management upgrades; Control and Display Unit 800Y (CDU-800Y) replacement; a new flight management system; and upgrades to the emergency locator transmitter, flight data recorder, mode 5 IFF, embedded GPS inertial with selective-availability antispoofing module/M-code GPS, attitude director indicator, horizontal situation indicator, attitude heading, reference system, flight director, VOR/ILS/Marker Beacon multimode receiver, and digital engine instruments.

Other improvements include a communications and networking upgrade, a multiphased effort that includes advancements in the Joint Tactical Radio System, integrated broadcast services, the family of advanced beyond-line-of-Sight Terminals, wideband line-of-sight and beyond-line-of-sight upgrades, advanced tactical data links integration, airborne networking, and network-centric operation enhancements.

After these improvements, the JSTARS program is now able to coordinate with and participate in projects developing international standards (including NATO standards) to ensure joint, allied, and coalition interoperability such as Attack Support Upgrade Link 16 enhancements, which have developed JSTARS into a controlling unit with full battle management capabilities.

In 2015, the E-8C JSTARS fleet was scheduled to be subjected to other modernization programs such as the "Kill Chain Enhancement Program," which monitors, identifies, evaluates, compares, and prioritizes projects that expediently deliver warfighting capabilities. The program focused on rapid implementation and delivery, rather than long-term production. The Air Force implemented emerging technologies that greatly increase system and system-of-systems capability, as well as interoperability with joint service, allied, and coalition systems. Representative efforts include imagery comparison, remotely piloted aircraft data integration, broadcast intel track correlation, multisensor radar service and tracker improvements, time-critical targeting initiatives, internet protocol–enabling technologies to enhance command and control to shorten the kill chain, machine-to-machine data exchange, enhanced targeting and interdiction, and radar and synthetic aperture radar enhancements.

The "JSTARS Modernization" is another upgrade program that encompassed multiple efforts to develop and integrate system improvements across the E-8C platform. Program accomplishments in fiscal year 2014 consisted of multiple efforts to develop and integrate platform-wide system improvements such as multi-agency communication capability, the Combined Enterprise Regional Information Exchange System, ground moving-target indicator risk reduction, and beyond-line-of-sight network architecture upgrades. Additionally, upgrade efforts to training and support systems, such as weapon systems trainer, the Navigator Training System, and mission crew trainers, included a mission maintenance trainer, a prime mission equipment-maintenance training device, and a mission system trainer.

In August 2017, it was announced the US Air Force had awarded Northrop Grumman Corporation a contract to upgrade existing radio terminals aboard the E-8C JSTARS fleet and replace them with Air Force Tactical Receive System–Ruggedized (AFTRS-R) terminals. The AFTRS-R assures capability for the JSTARS fleet and those interacting with the weapon system to receive intelligence reports, including threat warnings in hostile environments, ensuring undiminished battle management in support of warfighters in the air, on the ground, and at sea. The AFTRS-R provides data feeds from airborne and overhead electronics intelligence collectors and allows JSTARS to detect and track a host of mobile threats, including enemy air defense and theater ballistic missile assets. Its capability will modernize the Integrated Broadcast Service by replacing the current commander's tactical terminal / hybrid receive-only CTT/H-R radio. The modification also addresses cryptographic modernization and diminishing manufacturing source issues with the CTT/H-R radio.

The new radio is part of a program planned to augment the E-8C's operational capabilities, which also includes the Global Imagery Server, which allows for the display of worldwide imagery data on all JSTARS operator workstations, and the Automatic Identification System, which will provide JSTARS with a permanent, integrated solution for maritime identification of participating vessels.

Other upgrades already scheduled by the fiscal year 2017 budget included the Prime Mission Equipment (PME) Diminishing Manufacturing Sources (DMS), which is a top issue for fleet viability. PME is required for JSTARS to maintain net-centric war fighter capabilities—ground moving-target indicator (GMTI) and battle management command and control (BMC2) as specified in the Operational Requirements Document (ORD). The last major modification to the mission computing hardware took place during the Computer Replacement Program (CRP) from 1997 to 2005.

The obsolescence of parts, stemming from an overdue capital equipment replacement (more than twenty-year-old commercial off-the-shelf [COTS] equipment), requires a major modification to maintain the existing processing capabilities and specification compliance. Modification will address hardware and software DMS issues and COTS life cycle replacement for operator workstations (OWS), the central computing subsystem, and the Radar Airborne Signal Processor (RASP) subsystem. Modifications also include but are not limited to mission and maintenance crew trainers, software maintenance, and support systems at Robins Air Force Base.

It should be said that a "recapitalization" (or "recap" for short) program, planned to replace the E-8 JSTARS with another aircraft, has been funded and canceled a few times in recent years. But when this book was being finished, it was officially announced that such a program would be definitely canceled from fiscal year 2019 onward and that the JSTARS would receive upgrades to keep it operational until the 2020s, when the E-8 would be replaced for a more decentralized concept yet to be developed. The real status of the upgrades to be installed on the E-8 JSTARS as part of this initiative was not disclosed then. However, this final effort is linked to a series of upgrades, called the "Joint Stars System Improvement Program," divided into four parts, the latest of which, "Joint Stars System Improvement Program IV," is scheduled to last from October 2018 to October 2023. This program not only is planned to proceed with ongoing modifications but also comprises some new ones. Among the improvements scheduled to be installed aboard the JSTARS fleet are E-8C access to the Joint Worldwide Intelligence Communications System (JWICS), Avionics Diminishing Manufacturing Systems (DMS), primary mission equipment (PME-DMS), Sensor and Electronics Systems (SES), Beyond Line of Sight (BLOS), Windows 10 upgrade, Integrated Broadcast Service (IBS) modernization, SATURN, wideband satellite communications (SATCOM), CC refresh, Link 16 Mod, and crypto modernization, plus additional efforts.

In late November 2017, a twelve-month, $349.6 million contract for Total System Support Responsibility (TSSR) was signed between the US Air Force and Northrop Grumman Corporation. Under the JSTARS TSSR program, Northrop Grumman is partnered with the US Air Force Life Cycle Management Center to provide integrated logistics support to the 116th and 461st Air Control Wings for all facets of sustainment and support of the JSTARS fleet at Robins Air Force Base and forward-operating locations overseas.

A graphic demonstration of the updates to be installed in the AN/APY-7. *Northrop Grumman*

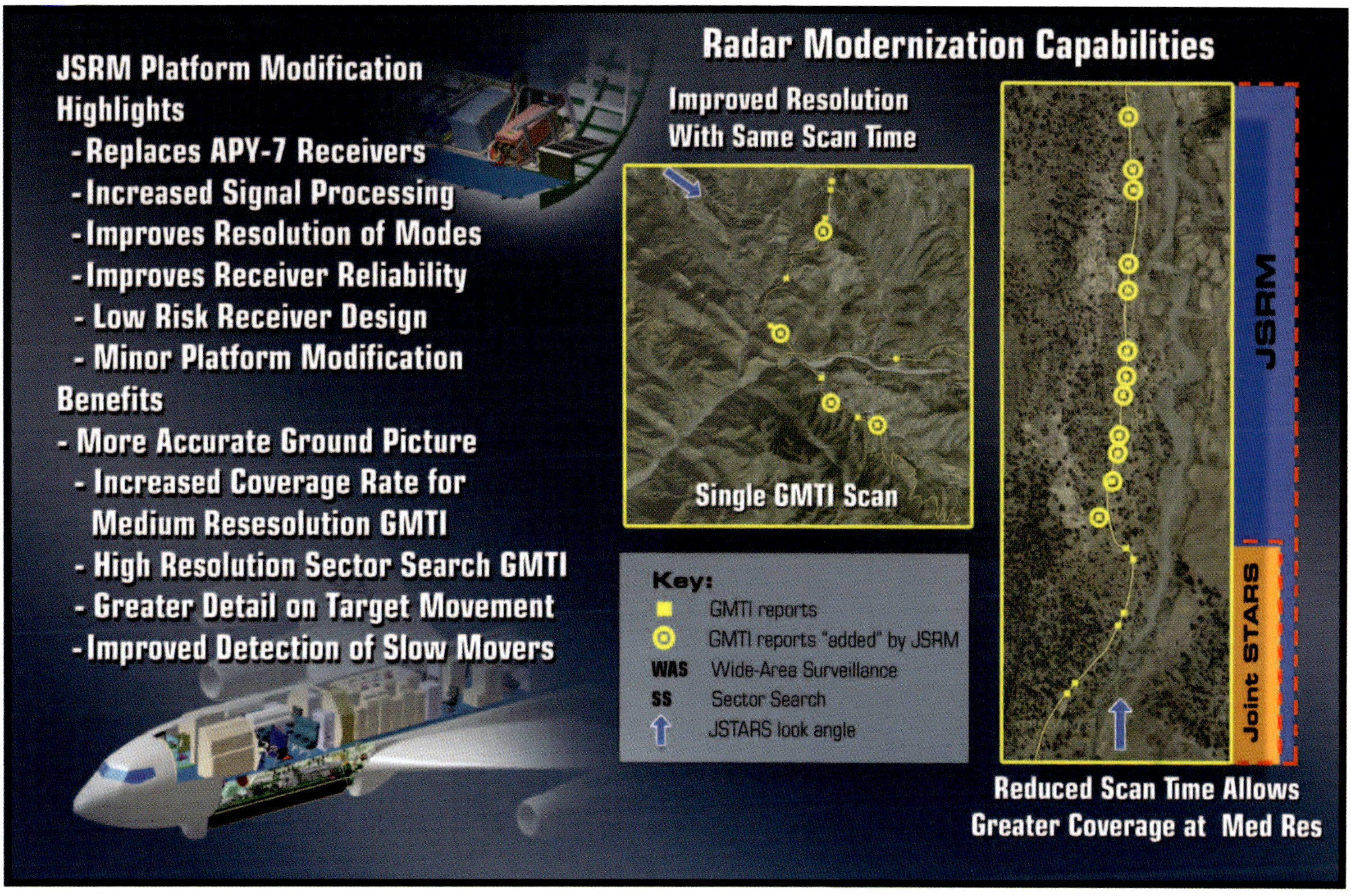

Another graphic demonstration of the updates planned to be incorporated into the AN/APY-7. *Northrop Grumman*

This program, along with the aforementioned JSSIP IV, will allow the E-8C to join the Advanced Battle Management and Surveillance, planned to be hosted at Robbins AFB. As per this initiative, three E-8Cs will be withdrawn from the active service in fiscal year 2019, followed by another example two years later. The remaining twelve aircraft will receive a comprehensive modernization package, made up of the Combined Enterprise Regional Information Exchange System, emergency locator transmitter, and common data link, in order to keep them in service until the late 2020s.

An overview of the capabilities to be incorporated into the E-8 after (and if) all the planned updates have been installed. *Northrop Grumman*

THE E-8C COMPARED WITH SIMILAR AIRCRAFT

Service Altitude

Raytheon ASTOR Sentinel R.1 42.000-50.000ft

Northrop Grumman E-8C 42.000ft

Embraer R-99 37.000ft

Shaanxi Y-8T 33.800ft

Tupolev Tu-214R 32.808ft

Operational Speed

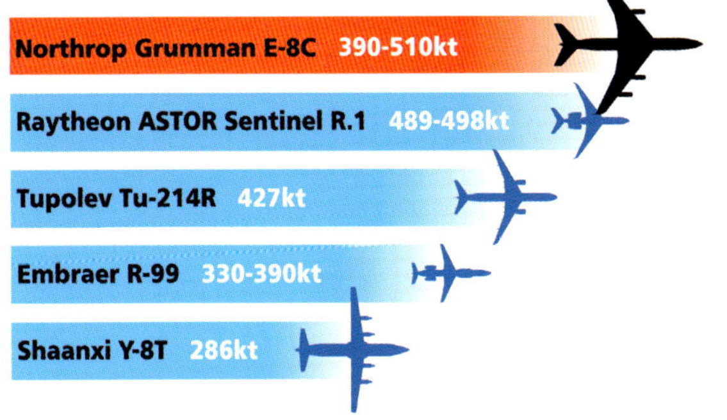

Endurance

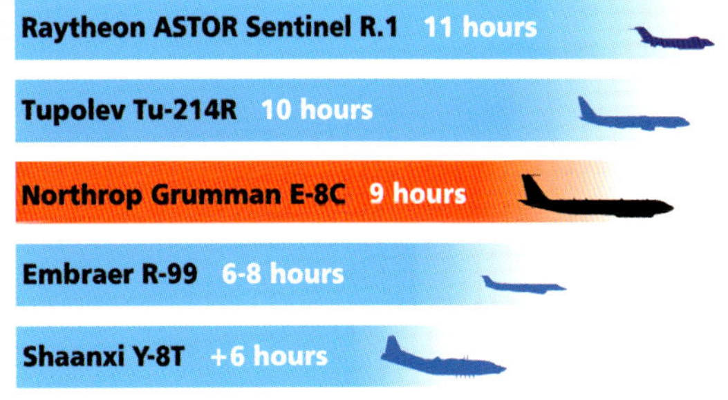

Radar Range / Radar Modes

186,4 miles

Raytheon ASTOR Sentinel R.1

Moving Target Search, Moving Target Spot, Stationary Target Search, Stationary Target Spot, Strip Mapping, Wide Area Spot.

+150 miles

Northrop Grumman E-8C

Synthetic, Aperture Radar/Fixed Target Indicator, Ground Moving Target Indicator.

150 miles

Embraer R-99

Air-to-Air
Spot SAR
Strip Mapping
Wide Area Spot
+21 others Modes

not avaliable

Tupolev Tu-214R

Ground Moving Target Indicator,
Ground Penetrating Radar.

not avaliable

Shaanxi Y-8T

Not Avaliable

CHAPTER 4

15th Wing (Provisional): Air Force Service Begins

The success obtained by the two E-8As during Operation Deep Strike was a decisive factor for their operational debut some months later. In August 1990, Iraqi forces invaded Kuwait, heightening a crisis that led the Western powers to concentrate their military forces mainly in Saudi Arabia to prevent a similar event against that country and push the invaders back to their territory. That concentration was called Operation Desert Shield, and some provisional units were created to command the aircraft for that theater of operations.

The 15th Air Division (Provisional or P), created on December 5, 1990, was one of those units responsible for reconnaissance and electronic warfare (EW) missions, for which it was equipped with RF-4C and F-4G Phantom IIs and EC-130H Compass Call EW aircraft. The E-8As and five IGSMs (soon augmented by another) were ordered to join it two weeks later, and as a result, two other "provisional" units were established: the JSTARS Operational Detachment One (made up of Army intelligence personnel to crew the IGSMs) and the 4411st JSTARS Squadron, under the 15th Air Division (P).

The E-8As were removed from their evaluation program and arrived in Riyadh, Saudi Arabia, on January 12, 1991. Their missions (to provide wide-area surveillance of the Allied area of operations, near-real-time radar imagery data of enemy movement, and possible targeting information) were controlled by the joint force air component commander (JFACC). All the radar information was relayed to the IGSMs based at the VII Corps HQ, the XVIII Airborne Corps HQ, the Army Central Command HQ, the Marines Central Command HQ, and the Tactical Air Control Center.

The JSTARS's first mission was flown on January 14, three days before the start of the offensive operations against Saddam Hussein's forces, called Operation Desert Storm. By the end of February, the E-8As had logged 535 flight hours in forty-nine missions with an operational availability exceeding 80 percent, thus being regarded as a vital air asset.

Side view of the USAF 60416. *Anderson Subtil*

Insignia of the very first unit to operate the JSTARS. *Anderson Subtil*

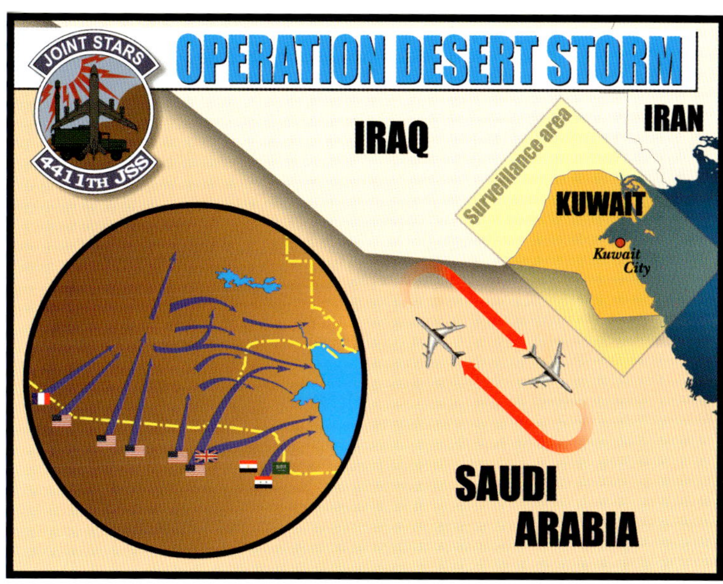

Map illustrating a typical E-8A mission during Operation Desert Storm. *Anderson Subtil*

CHAPTER 5

93rd Air Control Wing Service

In the early 1990s, multiple and opposite economic, political, and nationalist interests led to a breakup of Yugoslavia as a country, to the point that its former regions (Bosnia-Herzegovina, Croatia, Macedonia, Montenegro, Serbia, Slovenia, Kosovo, and Vojvodina) started not only to fight for their independence from Belgrade, the central power, but also to assume their predominance over each other.

As a consequence, a war began to escalate in August 1991. After three years, however, the Serbian hostile actions were stopped by a NATO intervention, which occurred again in 1995. By the end of that year, the "Former Warring Factions" (abbreviated FWF and made up of Serbian, Croatian, and Muslim contingents) were forced to adhere to the Dayton Peace Accord, which obliged them to withdraw from the adversary territories and gather heavy weapons at sites under NATO close surveillance. To prevent any violation of the agreement terms, the E-8 JSTARS were tasked with that mission; they would join with Operation Joint Endeavor as part of the United Nations Implementation Force (IFOR).

In order to be able to accomplish this assignment, an E-8A and one E-8C were removed from their evaluation program and prepared accordingly. They were operated by the 4500th Joint Surveillance Squadron (4500th JSS), a unit established for that operation, while the system's ground component was composed of personnel from the 303rd Intelligence Battalion, who crewed a mix of ten LGSMs/MGSMs. The stations were sent to Italy (San Vito, Combined Joint Special Operations Task Force; Aviano Air Force Base, GSM Task Force HQ; and Vicenza, Combined Air Operations Centre, CAOC), Germany (Rhein-Main Air Force Base, also the E-8 JSTARS Forward Operating Base), Hungary (Taszar, US Army in Europe [USAEUR] Forward Base; Kaposvar, home to an aviation unit), and Bosnia-Herzegovina (Gornji, Vokse, Sarajevo, Tuzla, Bieljla, and Vlasenica).

The JSTARS missions were planned by the CAOC, which had the intelligence/surveillance/reconnaissance cell design to provide near-real-time adaptations for a given sortie, according to air traffic management issues. Furthermore, the CAOC also had the Linked Operations Centre–Europe (LOCE), which disseminated and received NATO intelligence about the FWF.

The E-8/GSMs proved their usefulness when, for example, suspicious activity was detected near to the Sava River; the movement was later confirmed as executed by the FWF. After more than 1,000 combat hours distributed in ninety-five sorties, the JSTARS/GSMs were sent back to the US in December 1995.

However, by late September 1996, they were requested to be operational in Yugoslavia again. Their new task was to provide surveillance cover during the transition time from the IFOR to the SFOR (Stabilization Force), the "Operation Joint Guard." For that deployment, the E-8s, now under the redesignated 93rd Air Expeditionary Group, 93rd AEG, were based again at Rhein-Main AFB, while the GSMs were manned by the 319th Intelligence Battalion, grouped as "Task Force Dragon" and based in Mostar (French sector) and Banja Luca (British sector). The JSTARS's first eight-hour mission occurred in mid-November 1996. As was the case with Operation Joint Endeavor, the system's missions were managed by the CAOC in Vicenza. However, it is noteworthy that those missions over Yugoslavia did not run as smoothly as those against Iraq, mainly due to the mountainous terrain (causing masking and clutter issues yet to be fixed) and insufficient coordination within the command chain. As a result, poor operational results were obtained. Operation Joint Guard ended in June 1998.

An E-8C aircrew, mission crew, and maintainers pose with leadership of the 7th EACCS at Al Udeid Air Base, Qatar, on May 1, 2014, after reaching a milestone of 100,000 flying hours, most of which was flown during the previous missions. *US Air Force photo / Senior Airman Jared Trimarchi*

The aforementioned tendency showed by some of the former Yugoslavian provinces to try to prevail over others by force was the catalyst that unleashed another round of hostilities in the region. By late March 1999, this situation unleashed Operation Allied Force, an effort to stop the aggressions perpetrated by the FRY (Federal Republic of Yugoslavia) security forces against those of the KLA (Kosovo Liberation Army). For this operation, two E-8s were sent back to Rhein-Main AFB and flew under the direction of the CAOC in Vicenza. First they were tasked with supporting the air strikes (being part of the strike packages along with several types of aircraft), and then with the surveillance over the KEZ (Kosovo Engagement Zone, established some days after the hostilities began), as well as providing near-real-time intelligence and targeting information both to the CAOC and the EC-130E Airborne Battlefield Command and Control Center. However, whether the circumstances required avoiding collateral damages, collecting real-time intelligence, or neutralizing targets of opportunity, the E-8s could directly contact the forward air Controllers and vector NATO strikes.

However, as occurred in the previous operational missions flown by the E-8s over that theater of operations, its mountainous terrain impeded the system from detecting ground targets at certain angles, a fault compensated for by other assets in the order of battle, such as the U-2. Other factors against the JSTARS operational performances in that operation were that the enemy learned how to avoid its convoys being detected by the E-8s,

A frontal view of the same aircraft. *US Air Force photo / Senior Airman Jared Trimarchi*

by mixing them with civilian cars and refugee trucks, and as a result the JSTARS flew day missions only. Operation Allied Force finished on June, 10, 1999.

The terrorist attacks against the US on September 11, 2001, led to an immediate response by the Allied powers, called Operation Enduring Freedom, which officially commenced less than one month after those attacks, on October 7. The E-8s was authorized to augment the available ISR assets in early November. Their operations (which now also included, in addition to the detection of enemy ground forces, the airspace deconfliction and management of several simultaneous Close Air Support requests) were directed from a new CAOC, at Prince Sultan Air Base in Saudi Arabia, while the JSTARS themselves were based in Al Udeid, Qatar. The E-8 operated in concert with satellites and RQ-1 Predator UAVs and accumulated a good deal of the more than 1,300 flight hours logged by the several ISR aircraft in that theater of operations until December 2001, having been directly involved in the neutralization of more than 100 fixed targets, hundreds of vehicles or artillery pieces, and many concentrations of Taliban and al-Qaeda warriors located in the Afghan provinces. Operation Enduring Freedom in Afghanistan was officially terminated in December 2014, having been replaced by Operation Freedom's Sentinel, to be described later in this book.

A map depicting the area surveilled by the E-8s during Operation Allied Force. *Anderson Subtil*

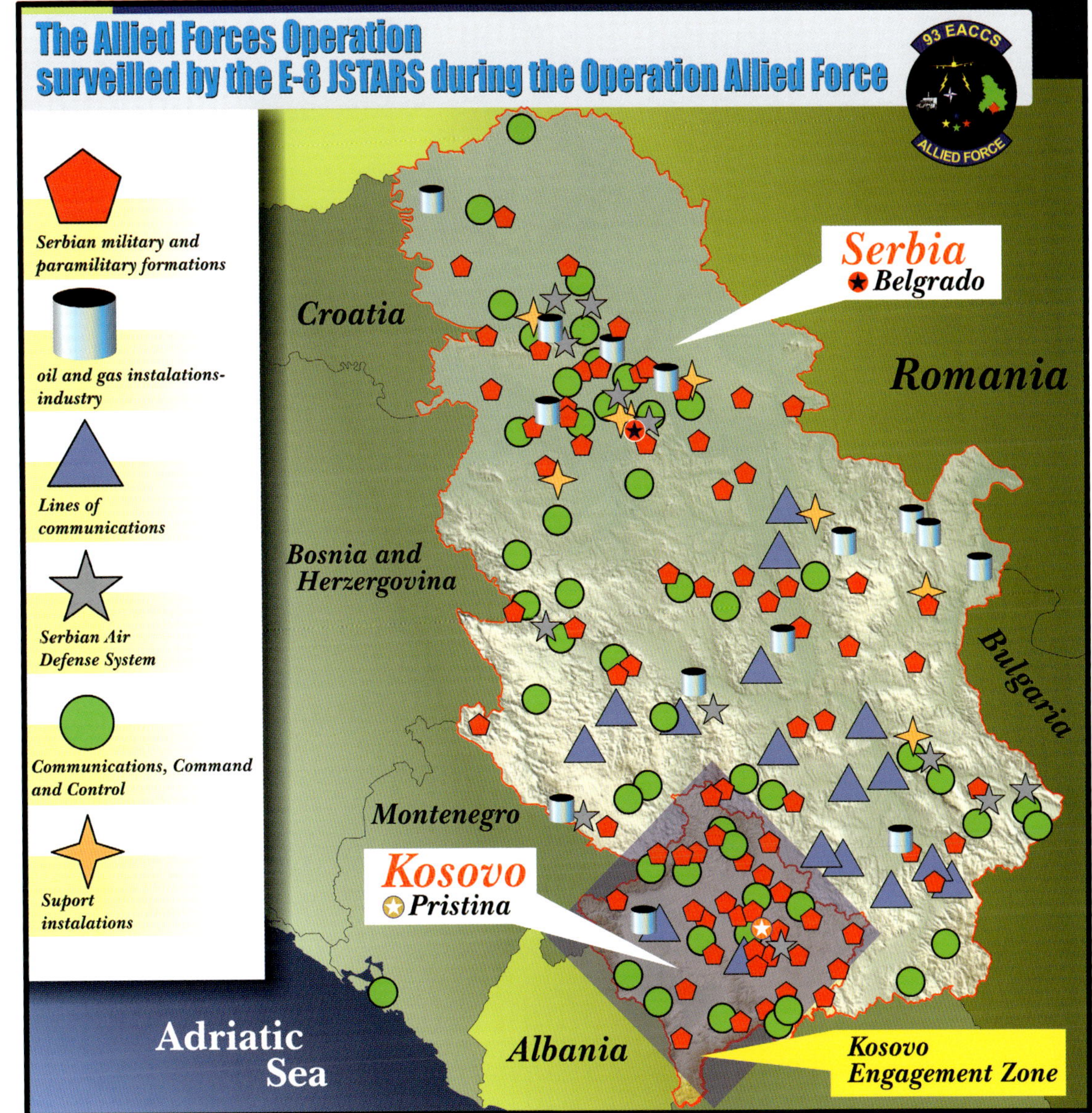

A map illustrating the Afghanistan territory during Operation Enduring Freedom. *Anderson Subtil*

A map with the zones surveilled by JSTARS missions during Operation Joint Endeavor. *Anderson Subtil*

The badge worn by the 93rd EACCS crews during Operation Allied Force. *Anderson Subtil*

Insignia of the 93rd Air Control Wing. *Anderson Subtil*

The badge worn by the 4500th JSS crews. *Anderson Subtil*

CHAPTER 6

116th Air Control Wing Service

On October 2, 2002, the 93rd ACW was deactivated and replaced by the 116th ACW, an Air National Guard unit. The wing continued to command the E-8s during Operation Enduring Freedom, but their aircraft were soon committed to a new deployment—Operation Iraqi Freedom, which commenced in March 2003 as a result of several accusations against the Iraqi government, including the support of terrorist groups.

This resulted in a US-led coalition invading Iraq. The aerial operations were managed by the same CAOC in Saudi Arabia, which now had a "time-sensitive target cell," planned to reduce the time required to neutralize critical threats such as surface-to-air missile mobile launchers. This operational structure directed the missions flown by the two E-8Cs (later augmented by another five or seven aircraft, according to some sources) in that conflict, where they performed vectoring of forward air controllers and strike aircraft against those targets. The JSTARS in Operation Iraqi Freedom, which ended in September 2010, were part of an international eighty-aircraft-strong armada specialized in ISR missions that amassed 42,000 battlefield images and 1,700 hours of moving-target indicators distributed in 1,000 sorties.

By same time, however, the operation changed its name and focus. It was retitled Operation New Dawn, planned to establish the transition of Iraq security forces from the US forces to the local ones, as well as to assist with the withdrawal of the US forces from Iraqi territory. Accordingly, the E-8s flew surveillance missions, during which they covered the removal of US convoys, protecting them against terrorist forces, until the operation ended in December 2011. Nine months earlier, however, the 116th ACW was committed to further operations.

The violent repression perpetrated by Colonel Muammar al-Gaddafi, the Libyan dictator, over the country's population fighting for democracy as a result of the so-called Arab Spring, a popular movement for freedom that commenced in Arab nations in 2010, prompted the international community to react. In early March 2011, as part of Operation Odyssey Dawn, US aircraft started to attack the Libyan military installations, but twenty days later the command of operations was forwarded to NATO. This effort was titled Operation Unified Protector. The E-8Cs were not involved in the first days of Operation Odyssey Dawn, obliging the fighter aircrews to assume the type's typical roles the best way they could; however, once present at the theater of operations, the E-8Cs flew eighty-four missions. For Operation Unified Protector, the JSTARS was stationed in Rota, Spain, while the CAOC for that effort was based at Poggio Renatico, Italy. The limited quantity of these aircraft (just one example was available) did not prevent them from providing offshore intelligence on enemy positions, although at a low rate. The JSTAR crews returned home in early November 2011, having contributed for the neutralization of over 5,900 military targets.

An E-8C Joint Surveillance Target Attack Radar System sits on a taxiway at Al Udeid Air Base, Qatar, on May 1, 2014. *US Air Force photo / Senior Airman Jared Trimarchi*

The E-8C Joint STARS, deployed to Al Udeid Air Base, Qatar, reached 90,000 combat hours here on July 23, 2014. Joint STARS crews have contributed tremendously to Operations Enduring Freedom, Iraqi Freedom, New Dawn, Odyssey Dawn, and Unified Protector. *US Air Force photo by SSgt. Ciara Wymbs*

A US Air Force E-8C Joint STARS from the 7th EACCS sits on the ramp before a mission over Iraq. This flight marks 40,000 combat hours for 116th Air Control Wing's Joint STARS supporting the Global War on Terror. *US Air Force photo / SSgt. Aaron Allmon II*

A trio of E-8Cs being prepared for the next mission. *USAF*

Team JSTARS aircraft crew chiefs recover an E-8C Joint STARS returning from a mission in support of Exercise Iron Dagger at Robins Air Force Base, Georgia, on June 15, 2012.
US ANG photo by MSgt. Roger Parsons

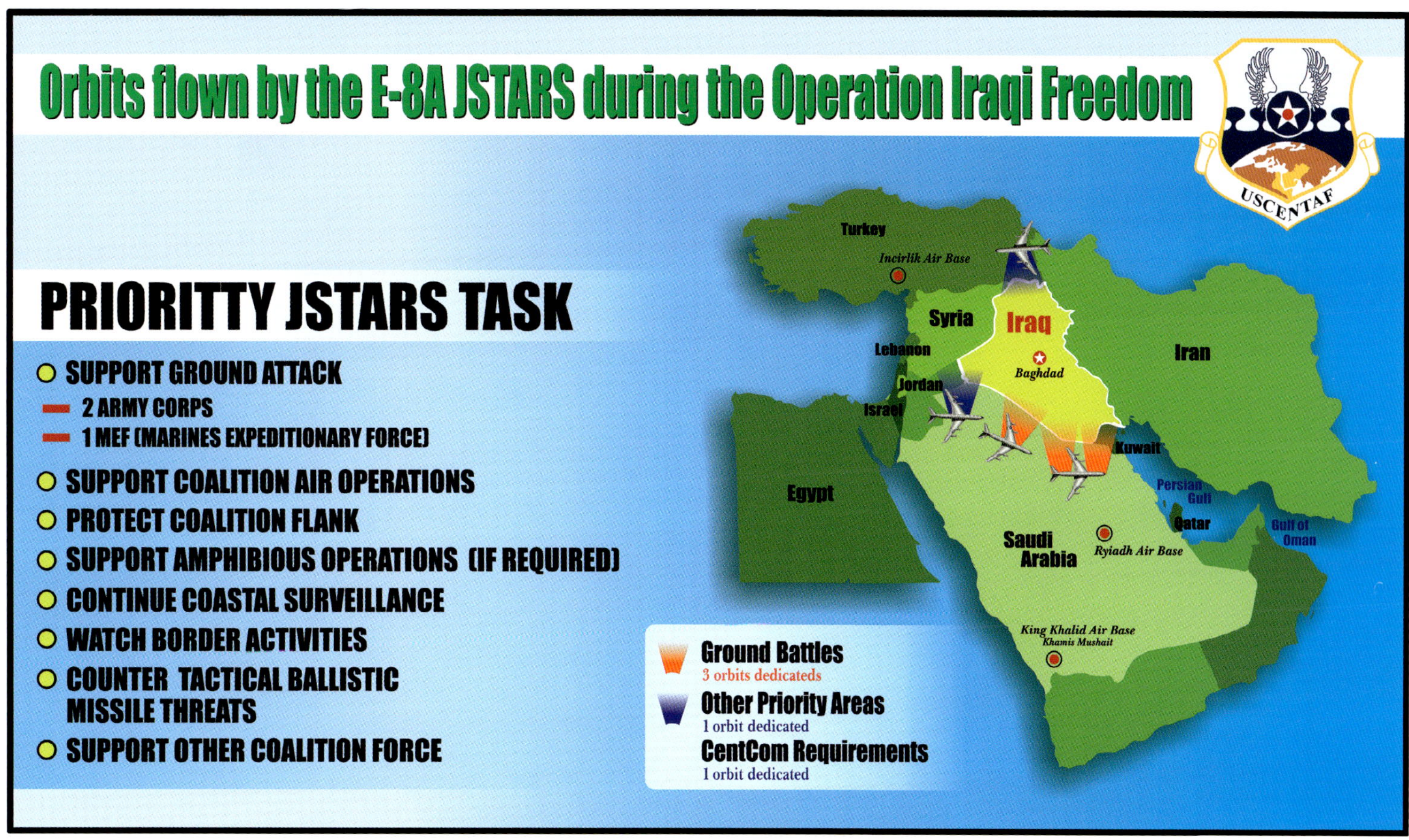

A map with the areas surveilled by the E-8Cs during Operation Iraq Freedom. *Anderson Subtil*

A map illustrating the areas where Coalition aircraft were based during Operation Unified Protector.
Anderson Subtil

The 116th Air Control Wing insignia. *Anderson Subtil*

The 116th Maintenance Squadron patch. *Anderson Subtil*

The 128th Airborne Command and Control Squadron patch. *Anderson Subtil*

CHAPTER 7

461st Air Control Wing Service

A few weeks prior to the official ending of Operation Unified Protector, a new E-8 unit was activated—the 461st Air Control Wing—as a joint USAF / US Army unit to share responsibility with the 116th ACW regarding the JSTARS operations.

The 138th Military Intelligence Company was the US Army's contribution to the new wing.

A US Air Force flight crew from the 116th Air Control Wing (ACW), Georgia Air National Guard, and the 461st ACW fly a night mission aboard an E-8C Joint STARS, Robins Air Force Base, Georgia, on July 13, 2017. *US ANG photo illustration by SMSgt. Roger Parsons*

Four E-8C Joint STARS aircraft sit on the flight line under the light of the moon at Robins Air Force Base, Georgia, on July 13, 2017. *US ANG photo by SMSgt. Roger Parsons*

An E-8C Joint STARS sits in a hangar during maintenance operations to repair hydraulic lines at Robins Air Force Base, Georgia, on July 13, 2017. *US ANG photo by SMSgt. Roger Parsons*

Crew chiefs from the 116th and 461st Air Control Wings catch an E-8 Joint STARS returning from a night mission at Robins Air Force Base, Georgia, on February 15, 2012. *US ANG photo by SMSgt. Roger Parsons*

A US Air Force E-8C Joint Surveillance Target Attack Radar System aircraft taxis down the runway during a morning mission at Robins Air Force Base, Georgia, on July 20, 2017.
US ANG photo by SMSgt. Roger Parsons

An E-8C test aircraft is displayed during a Joint Surveillance Target Attack Radar System exhibition in a hangar at Joint Base Langley-Eustis, Virginia, on December 18, 2012. *US Air Force photo by SSgt. Krystie Martinez / Released*

The same aircraft from a different angle. *US Air Force photo by SSgt. Krystie Martinez/Released*

An E-8C approaches the tanker with its refueling receptacle open. *USAF*

An E-8 JSTARS banks to the right, exposing its radar compartment. *Northrop Grumman*

The "Night Stalker" at dusk. *Northrop Grumman*

The 12th Air Command and Control Squadron patch. *Anderson Subtil*

The 16th Air Command and Control Squadron patch. *Anderson Subtil*

The 138th Military Intelligence Company patch. *Anderson Subtil*

The 461th Air Control Wing insignia. *Anderson Subtil*

The Air National Guard insignia. *Anderson Subtil*

CHAPTER 8

379th Expeditionary Air Wing Service

The 379th Expeditionary Air Wing was activated in April 2002 as a provisional unit, ready to be activated as an operational need if required. The 7th Expeditionary Airborne Command and Control Squadron (7 EACCS) is its operational unit, flying the E-8C JSTARS at the US Central Command Area of Responsibility, having been based at Al Udeid Air Base, Qatar, since March 2008.

In October 2014, the military actions against the ISIS (Islamic State of Iraq and Syria) terrorist group were called Operation Inherent Resolve, and the E-8Cs were essential tools to detect and track ISIS convoys throughout the Afghan terrain. A feature of this operation is that the targets of opportunity are found and strikes are coordinated in real time via communications and data links with the CAOC at Al Udeid AB.

In January 2015, Operation Enduring Freedom was officially replaced by Operation Freedom's Sentinel, planned to support counterterrorism operations against the remnant warriors of al-Qaeda based in Afghanistan in such a way that al-Qaeda will never again be able to conduct terrorist attacks against other nations.

This effort, as is the case with Operation Inherent Resolve, has also been carried out by the E-8s allocated to the 7th Expeditionary Airborne Command and Control Squadron, and continued at the time this book was being written, when more than 5,000 intelligence/surveillance/reconnaissance missions had been logged during 2017 only.

In addition to these operations, the 379th Air Expeditionary Wing flew some sorties to detect Islamic militants in Mali as part of Operation Serval, which lasted from January 2013 to July 2014 and was initiated by the French government after a formal request by the Malian interim government for military assistance.

An E-8C Joint Surveillance Target Attack Radar System aircraft takes to the sky over Al Udeid, Air Base, Qatar, on the morning of August 4, 2017, in support of Operations Inherent Resolve and Freedom's Sentinel. *US ANG photo by TSgt. Bradly A. Schneider*

Three E-8C aircraft sit on the ramp at Al Udeid, Air Base, Qatar, on the morning of August 4, 2017, as part of the efforts involved in Operations Inherent Resolve and Freedom's Sentinel. *US ANG photo by TSgt. Bradly A. Schneider*

US Air Force Senior Airmen Mehmet Yasdiman, *foreground*, and William Pryor, communications system technicians assigned to the 7th Expeditionary Airborne Command and Control Squadron, study their monitors aboard an E-8C aircraft out of Al Udeid, Air Base, Qatar, on July 27, 2017. *US ANG photo by TSgt. Bradly A. Schneider*

Portable oxygen tanks used in the event of an emergency are seen onboard an E-8C aircraft that launched out of Al Udeid, Air Base, Qatar, on July 27, 2017. *US ANG photo by TSgt. Bradly A. Schneider*

US Air Force Lt. Col. Ronnie Birge, mission crew commander assigned to the 7th Expeditionary Airborne Command and Control Squadron, studies his computer monitor during a mission aboard an E-8C aircraft out of Al Udeid, Air Base, Qatar, on July 27, 2017. *US ANG photo by TSgt. Bradly A. Schneider*

US Army Sgt. Joshua Smith, airborne target surveillance supervisor assigned to the 7th EACCS, studies his monitor aboard an E-8C aircraft out of Al Udeid, Air Base, Qatar, on July 27, 2017, in support of Operations Inherent Resolve and Freedom's Sentinel. *US ANG photo by TSgt. Bradly A. Schneider*

US Air Force Capt. Quincy Whitham, *foreground*, and 1st Lt. Elizabeth McLamb, air weapons officers assigned to the 7th EACCS, observe their monitors during a mission aboard an E-8C aircraft out of Al Udeid, Air Base, Qatar, on July 27, 2017. *US ANG photo by TSgt. Bradly A. Schneider*

US Air Force SSgt. Clea Herring (*foreground*), flight engineer assigned to the 7th EACCS, monitors the instrument panel aboard an E-8C aircraft during a mission out of Al Udeid, Air Base, Qatar, on July 27, 2017. *US ANG photo by TSgt. Bradly A. Schneider*

US Air Force Capt. Anthony DiBiase, combat systems officer (navigator) assigned to the 7th EACCS, sits at his station aboard an E-8C aircraft out of Al Udeid, Air Base, Qatar, on July 27, 2017. In this picture, one can see the E-8C's internal configuration. *US ANG photo by TSgt. Bradly A. Schneider*

US Air Force officers assigned to the 7th EACCS take part in a mission aboard an E-8C aircraft out of Al Udeid, Air Base, Qatar, on July 27, 2017. *US ANG photo by TSgt. Bradly A. Schneider*

US Air Force Capt. Erik Von Husen, aircraft commander assigned to the 7th EACCS, conducts a preflight inspection of an E-8C aircraft at Al Udeid, Air Base, Qatar, on July 27, 2017. *US ANG photo by TSgt. Bradly A. Schneider*

The 7th EACCS aircrew disembarking from a bus toward their E-8C prior to takeoff for another mission during Operation Inherent Resolve in June 2016. *US Air Force photo by Senior Airman Janelle Patiño*

An US Air Force KC-135 refuels an E-8C as part of the ongoing Operation Inherent Resolve, on September 22, 2017. *US Air Force photo by SSgt. Trevor McBride*

379th Expeditionary Air Wing crest. *Anderson Subtil*

7th Expeditionary Air Command and Control Squadron insignia. *Anderson Subtil*

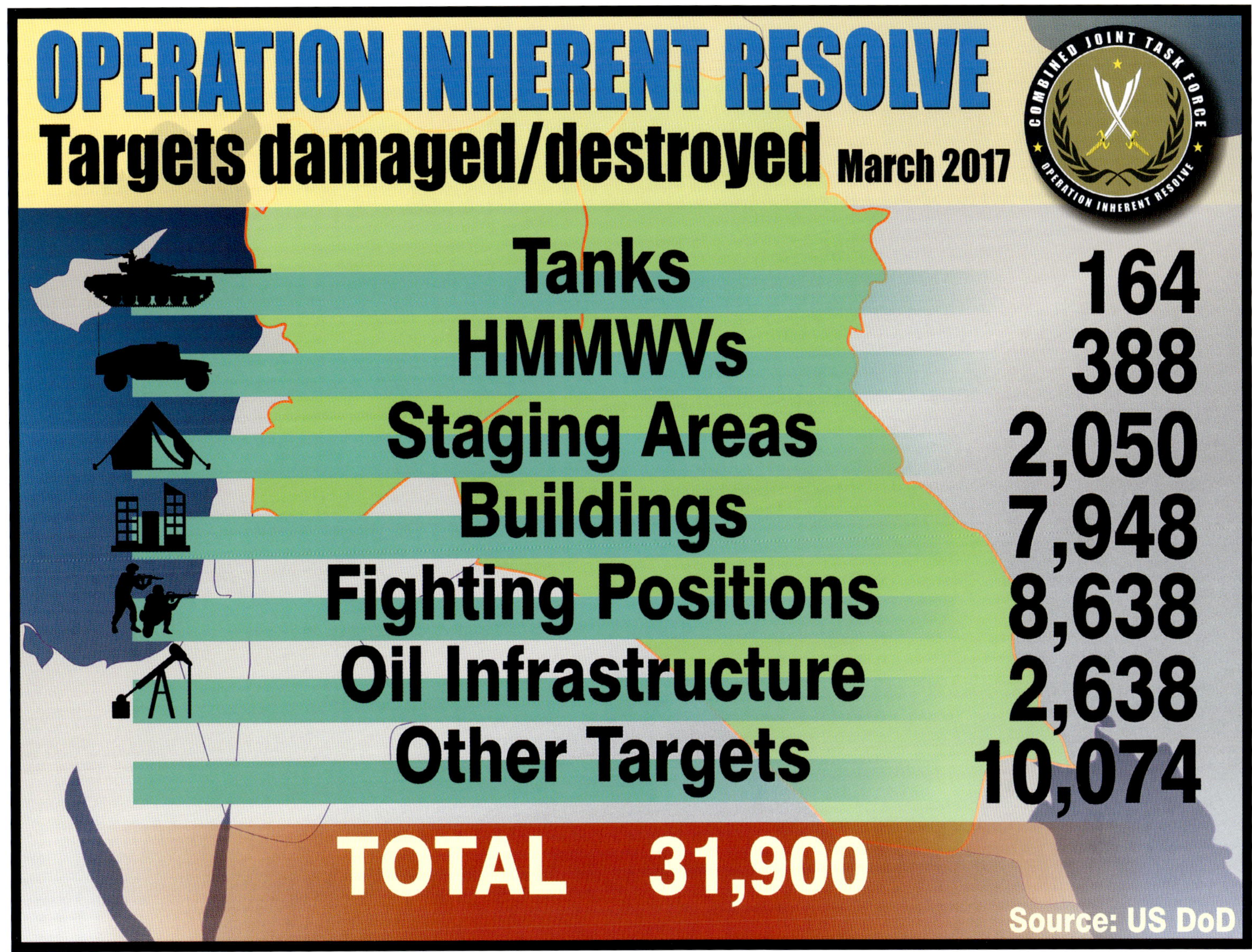

Operation Inherent Resolve results up to March 2017. *Anderson Subtil*

Glossary

ABCCC: Airborne Command and Control Center, a US Air Force aircraft equipped with communications, data link, and display equipment; it may be employed as an airborne command post or a communications and intelligence relay facility.

AC: Attack control, one of the operational modes of AN/APY-3.

ACW: Air Control Wing, an operational organization within the US Air Force. It comprises minor units, such as squadrons.

ADF: Automatic direction finder is an electronic aid to navigation that identifies the relative bearing of an aircraft from a radio beacon transmitting in the medium-frequency or low-frequency bandwidth, such as an NDB or commercial radio broadcast station. Essentially, ADF consists of a loop aerial that is rotated physically or electronically and detects the direction of minimum reception from the beacon relative to the aircraft direction. This information can be used directly to home in on the beacon; alternatively, the aircraft equipment combines the relative bearing with heading information from the aircraft compass to provide bearing position lines that may be plotted on a chart.

AEG: Air Expeditionary Group, an operational organization within the US Air Force, is a provisional wing/group concept used by the US Air Force. These units are activated under temporary orders by the owning MAJCOM for a specific purpose or mission. Once the subject mission is completed, these units are inactivated.

AESA: Active electronically scanned array, a phased-array antenna; that is, a computer-controlled array antenna in which the beam of radio waves can be electronically steered to point in different directions without moving the antenna. In the AESA, each antenna element is connected to a small solid-state transmit/receive module (TRM) under the control of a computer, which performs the functions of a transmitter or receiver for the antenna. This arrangement allows for quicker detection times to be obtained when compared with mechanically driven antennas.

AFTRS-R: Air Force Tactical Receive System–Ruggedized, a beyond-line-of-sight broadcast receiver for millions of threat, survivor, and Blue Force tracking reports daily without transmitting or the need for data forwarders. It is a multichannel, software-defined terminal that delivers Integrated Broadcast Service via the Common Interactive Broadcast waveform. It can use UHF SATCOM-secure voice, text, and data.

AIO/AIT: Airborne intelligence officer/technician, one of the specialists aboard the E-8C.

AIS: Automatic Identification System, a system that distinguishes ships by means of their name, course, speed, classification, call sign, or registration number. It can also provide specific data such as maneuvering information, closest point of approach, and time to closest point of approach and other navigation information. AIS transceivers automatically broadcast information at regular intervals via a VHF transmitter built into the transceiver. The information originates from the ship's navigational sensors, typically its global navigation satellite system receiver and gyrocompass.

ALARM: Airborne Long-Range Alerting Radar for MTI, one of the predecessors of the JSTARS concept.

AMSS: Airborne missions system specialists, stationed aboard the E-8C.

AOC: Air and Space Operations Center is a command center used by the US Air Force, being the senior agency of the Air Force component commander to provide command and control of air and space operations.

AOT: Airborne operation technicians, specialists aboard the E-8C.

AP: Attack planning, one of the operational modes of AN/APY-3 radar.

ATSC: Airborne target surveillance supervisors, specialists aboard the E-8C.

AWACS: Airborne Warning and Control System, the US designation for airborne early warning and control, a specific role performed by modified aircraft equipped to detect contacts at long ranges and direct forces against them.

BCE: Army Battle Coordination Element, an Army liaison team that operates at the US Air Force Tactical Air Control Center and provides a constant exchange of targeting information and intelligence that effectively integrates US Army operational requirements into the Air Tasking Order.

CAOC: Combined Air Operations Centre, a multinational headquarters for tactical and operational control of NATO Air Forces. Currently there are three CAOCs: Combined Air Operations Centre Uedem (CAOC UE) in Uedem, Germany; Combined Air Operations Centre Torrejon (CAOC TJ) in Torrejon, Spain; and Deployable Air Command and Control Centre (DACCC PR) in Poggio Renatico, Italy.

CGS: Common Ground Station, the E-8C JSTARS's ground mobile element.

CONUS: Continental United States, the area of the United States that is located on the North American continent. It includes forty-nine of the fifty states (excluding only Hawaii) and the District of Columbia, which contains the federal capital, Washington, DC.

CST: Communication system technicians, specialists aboard the E-8C.

DMCC: Deputy mission crew commander, specialist aboard the E-8C.

EACCS: Expeditionary Airborne Command and Control Squadron, a provisional unit dedicated to command and control roles to be activated if a needed.

ELT: Emergency locator transmitter, an integral component of the international satellite system for search and rescue (SAR) known as COSPAS-SARSAT. When activated manually—or automatically by immersion in water, or as a result of high g-forces on impact—ELTs transmit a distress signal at 406 and 121.5 MHz, which can be detected by nongeostationary satellites and then located precisely by GPS trilateration or Doppler triangulation or both.

FRY: Federal Republic of Yugoslavia, a federal state that existed from 1992 to 2003; formed by the republics of Serbia and Montenegro from the former Socialist Federal Republic of Yugoslavia (SFRY) after the other four republics broke state from Yugoslavia amid rising ethnic attacks, war, and tension.

FWF: Former Warring Factions, the armed forces that fought each other in the former Yugoslavia: Yugoslav Army (JA) and its predecessor, the Yugoslav People's Army (YPA), also referred to as the Yugoslav National Army (JNA); the Bosnian Serb Army (BSA); the Krajina Serb Army (SKA); the Croatian Army (HV); the Croatian Defence Council (HVO); and the Army of Bosnia-Herzegovina (BiH).

GRCA: Ground reference coverage area, one of the operational modes of AN/APY-3 radar.

GSM: Ground station module, one of the ground mobile elements of the E-8C JSTARS concept.

HDIS: High dynamics instrumentation set, one of the components of the GPS. It provides position, velocity, acceleration, attitude and attitude rates at ten times per second.

HF: High frequency, a frequency range between 3 and 30 MHz.

HQ: Headquarters.

IFF: Identification Friend or Foe, a two-channel system designed for identification of contacts, with one frequency (1030 MHz) used for the interrogating signals and another (1090 MHz) used for the reply. It has five modes: mode 1, which has sixty-four reply codes, is used in military air traffic control to determine what type of aircraft is answering or what type of mission it is on; mode 2, also only for military use, requests the "tail number" that identifies a particular aircraft—there are 4,096 possible reply codes in this mode; mode 3/A is the standard air traffic control mode and is used internationally in conjunction with the automatic altitude reporting mode (mode C) to provide positive control of all aircraft flying under Instrument Flight Rules; mode 4 is secure encrypted IFF, while mode 5 provides a cryptographically secured information to avoid collision and determine location.

IFOR: Implementation Force, a NATO-led multinational peace enforcement force in Bosnia and Herzegovina under a one-year mandate from December 1995 to December 1996, planned to implement the Dayton Peace Accords.

IGSM: Interim ground station module, one of the ground mobile elements of the E-8C JSTARS concept.

ILS: Instrument Landing System, a set of aids that allow landing operations during bad weather conditions. It is composed of a VHF localizer transmitter, an UHF glide slope transmitter, marker beacons, and an approach lighting system.

ISIS: Islamic State of Iraq and Syria, a militant jihadist terrorist organization and unrecognized proto-state that follows a fundamentalist, Wahhabi, and heterodox doctrine of Sunni Islam. It gained global prominence in early 2014, when it drove Iraqi government forces out of key cities in its Western Iraq offensive, followed by its capture of Mosul. ISIS originated as Jama'at al-Tawhid wal-Jihad in 1999, which pledged allegiance to al-Qaeda and participated in the Iraqi insurgency following Operation Iraqi Freedom. The group proclaimed itself a worldwide caliphate, and after successive military campaigns led by the United States and Russia, it was in decline by the time this book was being written.

ISR: Intelligence, surveillance, and reconnaissance, an activity that synchronizes and integrates the planning and operations of sensors, assets, processing, exploitation, and dissemination systems in direct support of current and future operations.

JFACC: Joint force air component commander, a commander within a unified command, subordinate unified command, or joint task force responsible for planning and coordinating air operations. The joint force air component commander is given the authority necessary to accomplish missions and tasks assigned by the establishing commander.

JFLCC: Joint force land component commander, a commander within a unified command, subordinate unified command, or joint task force responsible for planning and coordinating land operations. The joint force land component commander is given the authority necessary to accomplish missions and tasks assigned by the establishing commander.

JPO: JSTARS Joint Program Office.

JSTARS: Joint Surveillance Target Attack Radar System.

JTIDS: Joint Tactical Information Distribution System, also known as Link 16, is a military communications system that provides jam-resistant digital communication of data and voice. It works via a time division multiple-access network, with a maximum of 128 nets and 128 slots/sec./net, and has an information data rate from 28.8 to 238 KB/sec., with two channels of digital voice at 2.4 and 16 KB/sec. for cryptographic secure communications. The system's anti-jam features include fast frequency hopping, direct sequence spreading, and error detection and correction. Its maximum range is 500 nautical miles.

KEZ: Kosovo Engagement Zone, an area of Kosovo and southeastern Serbia where the Coalition aircraft were allowed to neutralize enemy contacts.

KLA: Kosovo Liberation Army, an ethnic-Albanian paramilitary organization that sought the separation of Kosovo from the Federal Republic of Yugoslavia (FRY) and Serbia.

LGSM: Light ground station module, one of the ground mobile elements of the JSTARS concept.

LOCE: Linked Operations Center–Europe, a system operated by the US European Command (USEUCOM) for imagery and information sharing between the US and its NATO allies; it includes communication resources and information systems and provides fairly robust email connectivity and web access between its user community and a number of designated NATO users and servers as well. LOCE also serves as a transport system for interconnecting constituent organizations.

LOS: Line of sight, when the radio transmitter and receiver are in the direct visual line of sight.

MCC: Mission crew commander, a specialist aboard the E-8C

MGSM: Medium ground station module, one of the ground mobile elements of the JSTARS concept.

MLRS: Multi-Lateration Radar Surveillance and Strike System, one of the predecessors of the JSTARS concept.

MTI: Moving-target indicator.

MTI-SLAR: Moving-Target Indicator Side-Looking Airborne Radar.

NATO: North Atlantic Treaty Organization, a military alliance created in 1949 by the United States, Canada, and several western European nations to provide collective security against the Soviet Union.

NLOS: Non-line-of-sight, when the radio transmitter and receiver are not in the direct visual line of sight, due to obstacles and long distances between them.

OB: Order of battle, an organizational and hierarchical disposition of a given armed force.

PEEK: Periodically Elevated Electronic Kibitzer, one of the predecessors of JSTARS concept.

PME: Prime mission equipment, a set of basic equipment needed for a given aircraft for perform its mission, such as radios and navigational aids.

RAP: Range Applications Program, a program dedicated to evaluate the GPS capabilities.

RRCA: Radar reference coverage area, an operational mode of the AN/APY-3 radar.

RSR: Radar Service Requests, operational needs issued by government agencies for the E-8C radar coverage to be executed over a given area.

SAR: Synthetic aperture radar, an airborne or spaceborne side-looking radar system that utilizes the flight path of the platform to simulate an extremely large antenna or aperture electronically, and that generates high-resolution remote-sensing imagery. Individual transmit/receive cycles are completed, with the data from each cycle being stored electronically. After a given number of cycles, the stored data are recombined to create a high-resolution image of the terrain being overflown.

SATC: Small area target classification, an operational mode of the AN/APY-3 radar.

SATCOM: Satellite communications.

SATURN: Second-Generation Antijam Tactical UHF Radio for NATO, a NATO-sponsored effort to develop a multinational antijam VHF/UHF radio covering the 225 to 400 MHz band with a single unit.

SCDL: Surveillance and control data link, a communication system that uses a secure, highly jam-resistant, dynamically alterable two-way digital data link for the control and distribution of information. The downlink transfers processed real-time radar imagery data from the airborne platform to a network of ground stations, while the uplink provides message communications between ground terminals and the airborne platform or between any two ground terminals in the net, using the airborne platform as a relay.

SD: Senior director, specialist aboard the E-8C.

SDS: Self-Defense Suite, a set of equipment designed to provide protection against threats directed to the E-8C.

SDT: Senior director technician, specialist aboard the E-8C.

SFOR: Stabilization Force, a military force deployed to stabilize the peace negotiated by the previous accords regarding Bosnia and Herzegovina.

SINCGARS: Single Channel Ground and Airborne Radio System, a combat radio network used by the US Armed Forces and some of their allies, capable of handling voice and data communication, both secure and nonsecure.

SMO: Sensor management officer, specialist aboard the E-8C.

SOTAS: Standoff Target Acquisition System, one of the predecessors of JSTARS.

TACAN: TACtical Air Navigation, an electronic navigation system, operating in the ultrahigh frequency (UHF) range. It is used by military aircraft to give the pilot both the distance and direction from the transmitting stations on the ground.

TADIL-J: Tactical Digital Information Link J, an improved and nodeless data link used to exchange near-real-time information. It is a communication, navigation, and identification system that supports information exchange among tactical command, control, communications, computers, and intelligence systems. The radio transmission and reception component of TADIL–J is the Joint Tactical Information Distribution System or its successor, the Multifunctional Information Distribution System. These high-capacity, ultra-high-frequency, line-of-sight, frequency-hopping data communications terminals provide secure, jam-resistant voice and digital-data exchange. JTIDS/MIDS terminals operate on the principle of time division multiple access, wherein time slots are allocated among all TADIL–J network participants for the transmission and reception of data.

TDA: Target Damage Assessment, the practice of assessing damage inflicted on a target from a standoff weapon, most typically a bomb or air-launched missile.

UAV: Unmanned aerial vehicle, an aircraft without a human being at the controls, remotely commanded from a fixed or mobile station.

UHF: Ultrahigh frequency, a frequency range between 300 MHz and 3 GHz.

USAEUR: US Army in Europe, a US Army major command, responsible for operations in fifty-one countries. It comprises commands, regiments, brigades, and groups.

UTM: Universal Transverse Mercator, a geographic and cartographic method used in aerial navigation, developed by Gerardus Mercator (1512–1594). The Universal Transverse Mercator system is a specialized application of the transverse Mercator projection. The globe is divided into sixty north and south zones, each spanning 6 ° of longitude. Each zone has its own central meridian. Zones 1N and 1S start at 180°W. The limits of each zone are 84° N and 80° S, with the division between north and south zones occurring at the equator.

VHF: Very high frequency, a frequency range between 30 and 300 MHz.

VOR: VHF omnidirectional range, with its frequency range standing between 108.0 and 117.95 MHz; every VOR is oriented to magnetic north (more on this in a bit), and emits 360 radials from the station. The VOR sends out one stationary master signal and one rotating variable signal. An aircraft's VOR antenna, which is usually located on the tail, picks up this signal and transfers it to the receiver in the cockpit. The aircraft's VOR receiver compares the difference between the VOR's variable and reference phase and determines the aircraft's bearing from the station. This bearing is the radial that the aircraft is currently on.

WAS: Wide-area surveillance, one of the operational modes of AN/APY-3 radar.

WD: Weapons directors, specialists aboard the E-8C.

Bibliography

Armistead, Edwin Leigh. *Airborne Early Warning Systems: An International Guide to Technologies, Projects and Markets*. London: SMI Publishing, 2000.

Brennan, Richard R., Jr., Charles P. Ries, Larry Hanauer, Ben Connable, Terrence K. Kelly, Michael J. McNerney, Stephanie Young, Jason Campbell, and K. Scott McMahon. *Ending the US War in Iraq: The Final Transition, Operational Maneuver, and Disestablishment of United States Forces–Iraq*. Santa Monica, CA: RAND, 2013.

Carey, Joel L. *Operation Odyssey Dawn and Lessons for the Future*. Research Report. Maxwell Air Force Base, AL: Air War College, 2013.

Cohen, Eliot A., dir. *Gulf War Air Power Survey*. Vol. 5, *A Statistical Compendium and Chronology*. Washington, DC: US Government Printing Office, 1993.

Commander's Handbook Distributed Common Ground System–Army (DCGS-A). Washington, DC: Department of the Army, 2009.

Cordesman, Anthony H. *The Air War against the Islamic State: The Need for an "Adequacy of Resources."* Washington, DC: Center for Strategic & International Studies, 2014.

Deptula, David A., Marc V. Schanz, and John M. Doyle. *Beyond JSTARS: Rethinking the Combined Airborne Battle Management and Ground Surveillance Mission*. Mitchell Institute Policy Papers 2. Arlington, VA: Mitchell Institute, September 2016.

Dondonis, Eduardo, Giovana Ester Zucatto, Tobias del Carvalho, Victor Merola, and Willian Moraes Roberto. "Combined Joint Task Force: Operation Inherent Resolve." *UFRGS Model United Nations* 3 (2015): 77–132.

Dunn, Richard J., III, Price T. Bingham, and Charles A. Fowler. *Ground Moving Target Indicator Radar and the Transformation of U.S. Warfighting*. Analysis Center Papers. El Segundo, CA: Northrop Grumman, 2004.

Fine, Glenn A., insp. *Operation Inherent Resolve*. Report to the United States Congress. Washington, DC: Department of Defense, 2017.

Gertler, Jeremiah, coord. *Operation Odyssey Dawn (Libya): Background and Issues for Congress*. Washington, DC: Congressional Research Service, 2011.

Greenleaf, Jason R. "The Air War in Libya." *Air & Space Power Journal* 27, no. 2 (March–April 2013): 28–54.

Gregory, Robert H. "Turning Point: Operation Allied Force and the Allure of Air Power." MMAS thesis. Ft. Leavenworth, KS: US Army Command and General Staff College, 2014.

Isby, David C. "JSTARS over Afghanistan." *Air Forces Monthly*, November 2010.

Lamb, Michael W. *Operation Allied Force Golden Nuggets for Future Campaigns*. Maxwell Air Force Base, AL: Air University Press, 2002.

Lambeth, Benjamin S. *Air Power against Terror: America's Conduct of Operation Enduring Freedom*. Santa Monica, CA: RAND, 2005.

Moseley, T. Michael. *Operation Iraqi Freedom: By the Numbers*. Shaw Air Force Base, SC: Assessment and Analysis Division, CENTAF-PSAB, 2003.

Phillips, Harry Vann. "Does the Joint Surveillance Target Attack Radar System (JSTARS) Support Military Peace Operations? A Case Study of JSTARS Support to Operation Joint Endeavor." Master's thesis. Ft. Leavenworth, KS: US Army Command and General Staff College, 1998.

Pun, A. K., W. R. Sheppard, and S. K. Purohit. *Corrosion and Fatigue Study of JSTARS Aircraft*. Vol. 1. El Segundo, CA: Military Aircraft Systems Division, Northrop Grumman, 1997.

Quintana, Elizabeth, and Jonathan Eyal, eds. *Inherently Unresolved: The Military Operation against ISIS*. Occasional Paper. London: Royal United Services Institute for Defence and Security Studies, 2015.

Royal Aeronautical Society. *Lessons Offered from the Libya Air Campaign*. Specialist Paper. London: Royal Aeronautical Society, 2012.

Rymer, Jon T., insp. *Operation Freedom's Sentinel*. Quarterly Report to the United States Congress. Washington, DC: Department of Defense, 2015.

So-Kargbo, Morleh, and Joshua McCarty. "Codifying Lessons Learned Teaches Something Fundamental about War." *Air Land Sea Bulletin*, January 2012.

Waldman, Andrew. "Eyes in the Sky." *National Guard*, May 2012.

E-8 JSTARS PRODUCTION

Construction	Previous	Serial	Conversion	Type	Notes
19626	N770JS	86-0416	FSD-1	E-8A	Former VH-EAH Qantas; initial designation EC-18C; delivered in August 1987; first flight December 22, 1988; European trials February–March 1990; Desert Storm 1991
19574	N8411	86-0417	FSD-2	E-8A	Former VH-EAF Qantas; initial designation EC18C; delivered in August 1988; first flight August 31, 1989; Exercise Deep Strike, September 1990; Desert Storm 1991; scrapped in 2000
24503	N707UM	88-0322	unknown	E-8B	First flight June 1990; Royal Saudi Air Force E-3A 1902
19621	N5265J	90-0175	FSD-3	E-8C	Former VH-EAA Qantas; TE-8C delivered in March 1994; Operation Joint Endeavor; named "Eye in the Sky"
19622	N4131G	92-1289	P-1	E-8C	Former VH-EAB Qantas; delivered in March 1996; Operation Joint Endeavor; named "Belle of Middle Georgia"
19295	N4115J	92-1290	P-2	E-8C	Former VH-EBV Qantas; delivered in December 1996; Operation Joint Endeavor
19294	67-30053	93-0597	P-3	E-8C	Former VH-EBU Qantas; delivered in November 1997; damaged beyond repair, 2009
19296	N6546L	93-1097	P-4	E-8C	Former VH-EBW Qantas; delivered in 1998; Operation Enduring Freedom
19293	N217BF	94-0284	P-5	E-8C	Former VH-EBT Qantas; delivered in 1999; Operation Enduring Freedom; named "Problem Child"
19442	67-30054	94-0285	P-6	E-8C	Former N370WA World; delivered in 1999; named "Radar Love"
20016	68-11174	95-0122	P-7	E-8C	Former JY-ADP Alia; delivered in 2000
20455	71-1841	95-0121	P-8	E-8C	Former N870PA Pan Am; delivered in 2000
20318	13705	96-0042	P-9	E-8C	Former Royal Canadian Air Force; delivered in April 1991; hit by Hurricane Rita (2005); returned to service
20316	13702	96-0043	P-10	E-8C	Former Royal Canadian Air Force; delivered in November 1997
190986	N707MB	97-0100	P-11	E-8C	Former F-BJCM Air France; delivered in August 2001
20317	13703	97-0200	P-12	E-8C	Former Royal Canadian Air Force; delivered in November 2001
20319	13704	97-0201	P-13	E-8C	Former Royal Canadian Air Force; delivered in April 2002
19998	10+02	99-0006	P-14	E-8C	Ex-*Luftwaffe*; delivered in August 2002
20043	85-6973	00.2000	P-15	E-8C	Former CF-ZYP Wardair; delivered in February 2003
19383	81-0892	Jan. 2005	P-16	E-8C	Former N7567A American Airlines; delivered in March 2004
19581	81-0896	Feb. 2011	P-17	E-8C	Former N8401 American Airlines; delivered in March 2005